26. 3. 79

The Physical Geography of the Tropics

The Physical Geography of the Tropics

An Introduction

John G. Lockwood

KUALA LUMPUR
OXFORD UNIVERSITY PRESS
OXFORD NEW YORK MELBOURNE

Oxford University Press

OXFORD LONDON GLASGOW
NEW YORK TORONTO MELBOURNE WELLINGTON
KUALA LUMPUR SINGAPORE JAKARTA HONG KONG TOKYO
DELHI BOMBAY CALCUTTA MADRAS KARACHI
IBADAN NAIROBI DAR ES SALAAM CAPE TOWN

ISBN 0 19 580319 1

Printed in Malaysia by Sun U Book Co. Sdn. Bhd., Kuala Lumpur
Published by Oxford University Press, 3, Jalan 13/3,
Petaling Jaya, Selangor, Malaysia

Preface

'THE Physical Geography of the Tropics' has been written for first-year University and as a reference for final-year school students in southern and eastern Asia, since the aim of the book is to introduce students to physical geography with particular reference to the tropics. Physical geography is often divided into a series of separate specialist sciences such as geomorphology, hydrology, climatology, pedology, etc., and introductory books tend to describe the elementary aspects of each of these sciences separately. This is unfortunate since it masks the inherent unity of the natural environment and leaves the student without an overall view of the subject. Instead, this book concentrates on those aspects of the subject that are important for an understanding of the natural environment as a whole. As the latter contains a number of complex systems, it is therefore necessary to introduce the subject with a brief dicussion of the nature of systems. Since the environment consists of complex circulation patterns of energy, water and matter as well as motion systems, separate sections are also devoted to these topics.

It is a formidable task to write about the physical geography of the tropics because a physical division of geographical area into tropical and non-tropical does not necessarily imply a similar division of the processes operating in these regions. Indeed, some of the processes of the temperate latitudes overlap into the tropical world and vice versa, and so many of the topics considered in this book have to be understood and appreciated by all physical geography students regardless of whether they live in tropical or temperate zones. Nevertheless there are important and distinctive differences between these climatic zones and these have been covered in this book.

My interest in the physical geography of the tropics is a long-standing one, and originated with studies of the climatology of India. Many people have helped to keep my interest alive, not least the staff and students of the Department of Geography, University of Malaya, Kuala Lumpur. In particular I would like to mention the late Robert Ho, one time Professor of Geography at the University of Malaya, for the encouragement he gave to my work while I was a member of his department.

The maps and diagrams in this book were drawn by Mr. T.P. Hadwin and Mr. G.S. Hodgson. The line drawings were made by Mr. G. Bryant. Lastly, I thank my wife for typing most of the manuscript and also for giving me great help and encouragement during its preparation.

Leeds
September 1974

JOHN G. LOCKWOOD

Acknowledgements

The author and publishers gratefully acknowledge permission received from the following to modify and make use of copyright material in diagrams and maps: the University of Chicago Press for Figure 4.1; the Elsevier Publishing Company for Figures 5.12 and 5.13; W.H. Freeman and Company (Scientific America, Inc.) for Figures 2.1 to 2.6; the McGraw-Hill Book Company for Figure 5.14; the New Zealand Meteorological Service for Figure 6.5; Oliver and Boyd for Figures 11.5 and 11.6; Prentice-Hall, Inc. for Figures 9.1 to 9.3; the Royal Meteorological Society for Figure 5.11; the Journal of Tropical Geography for Figures 6.2, 8.5, 10.2 to 10.4 and 11.1; the University Tutorial Press Ltd. for Figures 11.2 to 11.4; Springer-Verlag for Figure 4.9; George Weidenfeld and Nicolson Ltd. for Figures 4.3, 4.4, 6.3 and 6.4.

Full acknowledgements for photographs are given in the plate captions.

Contents

Preface   v   *Acknowledgements*   vi

PART A   INTRODUCTION

1 The Nature of Physical Geography   3   Peneplains   12
Method of Approach   3   Rocks   15
Systems   4   Igneous Rocks
    Isolated, Closed and Open Systems   Sedimentary Rocks
    Morphological, Cascading and   Metamorphic Rocks
        Process-response Systems   The Rock Cycle
Entropy   7   Geological Time-scales
Decaying, Cyclic and Haphazardly   Continental Drift and Plate Tectonics   16
    Fluctuating Systems   8   The Evolution of the Continents
    Decaying Systems   and Oceans   18
    Cyclic Systems   Present Structure of the Continents
    Haphazardly Fluctuating Systems   and Oceans   20
The Ecosystem   9   The Structure of South-East Asia   23
Tropical Ecosystems   9   The Structure of the Malay
Further Reading   10   Peninsula   24
                                                                 Highlands
                                                                 Old Alluvium
2 The Structure of the Earth   11   Modern Deposits
Isostasy   11   Further Reading   26

PART B   ENERGY IN THE TROPICAL ENVIRONMENT

3 The Nature of Energy   29   4 Radiation in the Tropical
Heat   29   Environment   36
Radiation   30   Short-wave Radiation   38
Potential Energy   30   Long-wave Radiation   42
Kinetic Energy   30   The Net Radiation   43
Chemical Energy   30   Energy Balance   44
Electric and Magnetic Energies   31   Bowen Ratio   47
Energy Transformations   31   Radiation within the Tropics   48
Energy in the Atmosphere   31   Further Reading   51
Bernoulli's Theorem   35

PART C MOTION AND MOTION SYSTEMS

5 Motion in the 'Natural Environment' 55
   Friction 56
   The Earth's Rotation 57
   Boundary Layers 59
   The Circulation of the Atmosphere 62
   Depressions and Anticyclones 66
   The Wind-driven Circulation of the Oceans 68

6 The Weather and Climate Systems
   of the Tropics 71

   Tropical Air Masses 71
   The Sub-tropical Anticyclones 72
   The Trade Winds 73
   The Equatorial Trough 75
   Rain-forming Disturbances 76
   The Monsoon Climates of
   Southern Asia 79
   The Climate of Equatorial South-East
   Asia 82
   Further Reading 85

PART D WATER IN THE TROPICAL ENVIRONMENT

7 The Hydrological Cycle 89
   Precipitation 92
   Evaporation and Evapotranspiration 95
   Interception 97
   Infiltration 99
   Soil Moisture 100
   Groundwater 101
   Water-balance 102

8 Rivers and River Systems 108
   Hydrographs 109
   Some Further River Hydraulics 113
   Floods 114
   Transportation, Deposition and
   Erosion by Rivers 118
   Further Reading 119

9 Glaciers and Glaciation 120
   The Elementary Physics of Ice and Snow 121

   Glacial Erosion and Deposition 123
      Striations
      Roches Moutonnees
      Drumlins
      U-shaped Valleys
      Cirques
      Moraines
      Kames
      Kettles
      Eskers
      Valves

   The Present-day Distribution of
   Ice-sheets and Glaciers 125
   Pleistocene Glaciations 127
   Wurm Climates 128
   Glacial—Eustatic and Isostatic
   Controls of Sea-level 129
   Further Reading 130

PART E THE CHEMISTRY AND BIOLOGY OF THE TROPICAL ENVIRONMENT

10 Matter in the Tropical Environment 133
   The Nature of Matter 133
   The Solvent Power of Water 134
   Chemical Cycles 135
   Ocean Water and River Water 135
   Weathering 136
      Physical Weathering
      Chemical Weathering
      Common Weathering Products
      Weathering and Climatic Conditions
   The Carbon Dioxide—Calcium
   Carbonate System 138
   Erosion 140
   Landforms of the Malay Peninsula 145

11 Landscape Ecology and Tropical
   Ecosystems 147
   Soils 151
      Soil Horizons
      Soils of the Humid Tropics
      World Distribution
      Desert Soils

   Vegetation and Ecosystems 158
      Tropical Rain Forest
      Vegetation Succession
   Further Reading 160

Index 161

PART A
Introduction

Volcanoes are important landscape elements in some parts of
tropical Asia. There are about 11 active volcanoes in the
Philippines and some 77 in Indonesia.

1 The Nature of Physical Geography

Method of Approach

SINCE the purpose of physical geography as a discipline is to study the natural environment, it is concerned primarily with the land surface, oceans, atmosphere, soil and vegetation mantle of the earth. However, physical geographers are not just interested in the natural environment for its own sake, but also because it forms the environment in which man lives. Thus the emphasis is rather different in this discipline from that of the traditional geophysical sciences in that it stresses those aspects of the natural world which affect man's activities.

There is an unfortunate tendency to divide the study of physical geography into separate specialist sciences such as geomorphology, hydrology, climatology, pedology, etc., and to consider each component as if it were completely self-contained. Though this has merit at an advanced level of study, it tends to lead to an under-valuation of the importance of the interactions and cycles which extend throughout the natural environment. Therefore, this traditional division of physical geography into separate specialist sciences has not been followed in this book, but the division is rather on the basis of the type of interaction or cycle involved.

The particular view that is taken of the natural environment depends on the space and time-scales chosen. Rates of change vary considerably, and so on a long geological time-scale the whole surface of the earth is in a state of continual flux and nothing is permanent, but on the time-scale of human existence many geological features of the earth's surface may be considered as almost fixed and permanent features of the landscape. Other aspects, such as plant growth, have annual or even shorter cycles, and therefore they may vary many times during a normal human life.

Since physical geography is largely concerned with time-scales ranging up to several thousand years, some

major features of the landscape may be judged to be fixed, while others go through many cycles. The major short-term cycles arise from the diurnal and annual variations in solar radiation, which either directly or indirectly control energy inputs, atmospheric circulation, rainfall, plant growth, etc. Environmentally, the relatively short-term cycles in energy, moisture and chemicals are of great importance, since they determine the general aspect of the natural environment—hot or cold, wet or dry—and should therefore form an important part of any general examination of physical geography.

Quasi-permanent features of the landscape such as hills, valleys, seas, etc. may be regarded only as the background scenery on a stage where a play is being performed in which the principal actors are the short-term cycles of energy, water and chemicals which interact with one another to produce the natural environment. Hence physical geography is considerably more than the study of the topographical landscape. Indeed the actual topographical landscape is not of great environmental importance in itself, since greatly differing environments can be found in similar topographical settings, e.g. hilly country in dry and humid areas. The short-term variations of energy, water and chemical flows can be difficult to understand because they are not immediately observable in the same way that a hill or river terrace is observable. Partly because of this and also because of a certain philosophical outlook, there was a tendency in the past in physical geography to stress the study of the historical evolution of landscapes, without bothering to ascertain the nature of the processes taking place at the actual time of study. As will be shown later there is little point in trying to describe the historical evolution of a landscape without any detailed understanding of the cycles of erosion and sedimentation taking place at the present time. This comment does not imply criticism of the work of earlier physical geographers, since other subjects have gone through a similar general descriptive stage before becoming quantified into exact sciences—a process which is now occurring in the study of physical geography.

Biological aspects of the natural environment have not so far been mentioned in this discussion of physical geography, since it is assumed that plant growth cycles and distributions are very much controlled by the cycles of energy, water and chemicals. In the zoological world it is possible to observe very complex food chains, for instance, where very small creatures such as insects are eaten by large creatures like birds which in turn die and are eaten by beetles, etc. Thus the same chemical substances are used by several creatures and a break at one point in the chain could lead to a lack of food for all creatures beyond that point; similar chains can be observed for the use of energy and water. These very complex chains and interactions which extend throughout both the physical and biological parts of the natural environment are often best described in terms of various types of systems or related sub-systems.

The application of both systems theory and mathematics to physical geography has changed the subject completely in recent years. The quantification of an otherwise rather descriptive subject arises basically because of a general feeling among physical geographers which was expressed very well many years ago by the distinguished Victorian scientist Lord Kelvin when he wrote, 'When you can measure what you are speaking about and express it in numbers, you know something about it; but when you cannot measure it, when you cannot express it in numbers, your knowledge is of a meagre, unsatisfactory kind.'

Chorley and Kennedy (1971) have powerfully stated the case for the use of the systems approach in physical geography. The use of systems and models enables the immensely complex real world to be examined and thought about in an orderly, logical manner. Associated with the growth of this systems approach has been a change from the descriptive study of the history of landscapes to a growing interest in the interactions and exchanges which shape both the landscape and the physical environment. Since the interactions consist of flows of energy and matter, much of the recent work in physical geography has concentrated on tracing these flows. Understanding energy and matter flows involves a knowledge of not only geomorphology and climatology, but also of the associated soils, plants and animals, and the study of all these is usually integrated under the general heading of the ecosystem. Thus modern physical geography is very much concerned with the study of ecosystems and their interactions with the socio-economic systems of human geography, and it is this interest which separates it from physical geology or the earth sciences.

Systems

A system may be defined as a structured set of objects and/or attributes, where these objects and attributes consist of components or variables that exhibit discernible relationships with one another and operate together as a complex whole, according to some observed pattern. The concept of the system is very useful in providing a means of understanding complex phenomena, provided that it is clearly understood that systems try to describe what happens in nature, and that nature cannot necessarily be forced into the mould of some particular preconceived system. Systems can be classified in terms of their function and also in terms of their internal complexity.

ISOLATED, CLOSED AND OPEN SYSTEMS

A common functional division is into isolated, closed and open systems.

(a) Isolated systems have boundaries which are closed to the import and export of both mass and energy. Such systems are rare in the real world, though they may occur in the laboratory, i.e. a mass of gas within a completely sealed and insulated container.

(b) A closed system is one in which there is no exchange of matter between the system and its environment though there is, in general, an exchange of energy. The planet earth together with its atmosphere may, very nearly, be considered a closed system.

(c) An open system is one in which there is an exchange of both matter and energy between the system and its environment. There are numerous examples of open systems in nature, i.e. precipitating clouds, river catchments, plants, etc.

Isolated Systems. Gas within a completely sealed and insulated container provides a good example of an isolated system. Whatever the original temperature gradients within the gas, temperatures will eventually become uniform, and while the system remains isolated nothing can check or hinder this inevitable levelling down of differences. Stated more generally, in an isolated system there is a tendency for the levelling down of existing differentiation within the system, and towards the progressive destruction of the existing order. In such a system there is always a decrease in the amount of free energy available for causing changes and doing work, and eventually the free energy will become zero. The classical view of landscape development includes the idea of an initial uplift providing a given amount of potential energy and that, as degradation proceeds, the energy of the system decreases until at the stage of peneplanation there is a minimum amount of free energy as a result of the levelling down of topographic differences. This view of landscape development seems to assume that the landscape acts as an isolated system. Furthermore, the classical belief in the sequential development of landforms, involving the progressive and irreversible evolution of almost every facet of the landscape with the reduction of relief, is in accord with the behaviour of isolated systems. The condition of an isolated system at any particular time can be considered to be largely a function of the initial system conditions and the amount of time which has subsequently elapsed. Thus isolated systems are very suitable to study on a historical basis. This suggests further analogies between the philosophy of isolated systems and the historical approach to landform study which was used by the early geomorphologists.

Open Systems. Open systems need an energy supply for their maintenance and preservation, and are in effect maintained by the constant supply and removal of material and energy. Closed systems may be considered as a special case of open systems, there being no exchange of matter with the environment. It has already been noted that most of the systems observed within the natural environment belong to the open group. In particular, the open system has one important property which is not found in the isolated system, that is, it may attain a condition known as steady-state equilibrium. This is the condition of an open system wherein its properties are invariant when considered with reference to a given time-scale, but within which its instantaneous condition may oscillate due to the presence of interacting variables. Stated rather more simply, the general features of the system appear to remain constant over a long period of time, though there may be minor changes in details.

The consequences of a steady-state condition can be illustrated by reference to an astronomical example. A galaxy is a spiral system of stars, such as the Milky Way to which our sun belongs, containing perhaps ten thousand million (10^9) stars. Now it is observed by astronomers that all galaxies are receding from us, with those farthest away moving fastest. This does not imply that we are at the centre of the universe, for the same result applies regardless of where the observer is situated in the universe. The easiest way to explain this expansion is to assume that at some time in the past, all the matter in the universe was present in one fantastically dense ball. This 'primeval atom' as it was called, exploded with cosmic violence, and the pieces condensed to form the stars and galaxies. The receding galaxies still represent the remains of this original explosion. This is the so-called 'big bang' or evolutionary theory of the universe and it contains many elements of isolated system philosophy. The universe has a definite beginning in time, after which it evolved with a gradually decreasing average density of matter. Thus the universe as a whole has a history and an observer would be able to tell the age of the universe by observing the average density of galaxies. Presumably such a universe would have some sort of ending.

Another theory of the universe states that as the galaxies recede from one another, new matter, in the form of hydrogen atoms, appears in the emptying spaces in just sufficient quantity to maintain the average density of matter at a constant value. The new matter eventually coalesces into stars and galaxies, and as the galaxies get further away they eventually recede from sight. The major features of such a universe will always appear the same to an observer since the universe is in a steady state. This theory of the universe, therefore, is known as the steady-state theory. By such a view the universe as a whole always looks the same, has no beginning, nor will it have an end. Observation of the universe as a whole cannot possibly reveal its age. Such a universe is an open system in a steady state.

Since it is important for astronomers to decide which of these two views of the universe is correct, much recent astronomical research has been directed towards this particular end. This is also an example of how the observer's basic philosophy might well influence what he expects to see in the natural world. Many segments of the natural environment such as individual slopes, stream catchments, plant communities, etc., form open systems which are in a steady state. They would therefore maintain their form and structure over long periods of time, for example, an erosional slope would retain its angle of repose despite the progressive removal of mass with time. Furthermore, once a steady state has been established, the influence of the initial conditions vanish and, with it, the evidence for a previous history of the system. Thus the only way to understand the operation of an open system in a steady state is to consider the interactions taking place within the system at the present time. Recent glaciations have violently changed middle-latitude landscapes, so few large-scale natural systems have had time to reach a steady state and the landscape contains many elements which form evidence of previous events. In contrast, within the tropics conditions have been relatively stable for long periods of time and many natural systems may be in a steady state.

A further characteristic of an open system is that free energy can be imported into it, stopping the trend towards uniformity that is found within isolated systems. Given large imports of free energy, open systems may even develop toward higher and more complex orders of differentiation and organization. Therefore, within open systems extreme complexity may not be a sign of youth but of great age.

MORPHOLOGICAL, CASCADING AND PROCESS-RESPONSE SYSTEMS

Systems may also be classified (Figure 1.1) according to their internal complexity, and one useful classification in physical geography is into morphological systems, cascading systems, and process-response systems.

Morphological Systems. Morphological systems are made up of the morphological or formal instantaneous physical properties integrated to form a recognizable operational part of physical reality, the strength and direction of their connectivity being commonly revealed by correlation analysis. For example, the morphological properties of a slope might include such parameters as maximum and average angle of slope, length of slope, depth of soil, amount of soil moisture, percentage vegetation cover, etc. The relationship between these various parameters is ex-

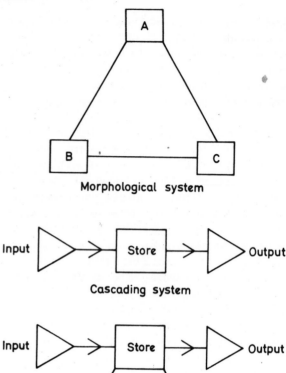

Figure 1.1 Schematic illustrations of three types of systems classified according to their internal complexity.
Morphological system. A system identified on the basis of the existence of significant correlations between its morphological properties.
Cascading system. A system made up of a chain of subsystems, each having both spatial magnitude and geographical location, which are linked dynamically by a cascade of mass or energy.
Process-response system. A system formed by the intersection of morphological and cascading systems.

pressed by a series of correlation coefficients, and the operational efficiency of the system is interpreted in terms of the degree to which these morphological parameters are related. Thus a morphological system is identified on the basis of the existence of significant correlations between its morphological properties.

Cascading Systems. Cascading systems are composed of a chain of sub-systems, having both magnitude and geographical location, which are dynamically linked by a cascade of mass or energy and in this way, the output of mass or energy from one sub-system becomes the input for the next sub-system. Typically, sub-systems consist of an input into a store, which may contain a regulator controlling the amount of mass or energy remaining in the store or forming the output. The regulator may be a physical property of the store itself or it may be completely external to the store. Also it may operate ahead, within or behind the store. More complex sub-systems may have several inputs and outputs and even several regulators which decide how the mass or energy is divided between the various outputs. Many of the processes taking place in the natural environment can be interpreted in terms of cascading systems, a good example being provided by the cycle of water. Water may be stored in the oceans, the atmosphere (as water-vapour), the soil, the deep rocks, rivers, etc., and the transfer of water from one store to another is controlled by various physical regulators. The output from the atmospheric store in the form of rain constitutes the input into the soil, where in turn one of the outputs forms the input into the deep rock storage, and so on until the water arrives back into the ocean where evaporation forms the input into the atmospheric store.

Interception of rainfall by tropical forest is a good example of a sub-system (see Figure 7.4). The amount of water that can be carried on a leaf surface is limited, and so there is a definite upper limit to the amount of water that can be stored in a tree canopy and thus to the store of the sub-system. The input into the sub-system is rainfall and the outputs are the evaporation of the intercepted water and the gradual drip of water out of the trees onto the soil surface. At the start of the rainfall the tree canopies will be dry and no water will reach the soil, but after some time the canopies will become completely saturated with water, and when this occurs most of the succeeding rainfall will eventually dip onto the soil surface. So the regulator controlling the amount of water reaching the soil surface will be the physical geometry of the tree canopies and the percentage saturation of the canopies. There is also a loss of water by evaporation from the intercepted water in the canopies. This loss is controlled by the prevailing meteorological conditions and thus by a regulator which is outside of the physical bounds of the sub-system.

Cascading systems can be extremely complex and often the details of the systems are not completely understood. Under these conditions great emphasis is often placed on the statistical relationships between the inputs and outputs, the actual internal structure of the system being ignored. When this method of analysing a cascading system is used it is said that the system is treated as a black box. In contrast, if an attempt is made to identify and analyse as many of the storages, flows and regulators within the system as possible, then it is said that the system is treated as a white box. The conversion of rainfall into river-flow is another example of a cascading system. It is possible to produce a series of equations which relate, for a given river, the river-flow to the rainfall, and this can be done without any knowledge of the physical processes taking place within the catchment. Such a series of equations form a black box model of the catchment, and they give no information on the actual physical processes involved in the conversion of rainfall into river-flow. If all the sub-systems within the catchment are carefully identified and examined then it is possible to produce a white box model of the catchment.

Process-Response Systems. Process-response systems are formed by the intersection of morphological and cascading systems, the links between them being commonly provided by morphological components which either coincide with, or are closely correlated with, storages or regulators embedded in the cascading system. For example, infiltration capacity and soil moisture may be both morphological properties of a slope system and also a regulator and a store respectively in a catchment cascading system.

Entropy

In very general terms, entropy may be defined as a measure of the disorganization in a system organization, that is to say the probability of encountering given states, events or energy levels throughout the system. Entropy is said to be at a maximum when there is an equal probability of encountering given conditions and this occurs when there is a complete disorganization within the system. Similarly, minimum entropy corresponds to a high degree of organization and hierarchical structuring.

Within an isolated system, energy is continuously redistributed throughout the system and the degree of disorganization increases with time. So after the initial formation of an isolated system, when entropy will be at a minimum and differentiation at a maximum, entropy will increase until it reaches a maximum value when the system is completely undifferentiated. In contrast, open systems continuously import energy and therefore can maintain or even increase their internal differentiation and hierarchical organization. An open system does not therefore tend towards a condition of maximum entropy, and characteristically has paths of flow of energy and matter.

Decaying, Cyclic and Haphazardly Fluctuating Systems

Open systems in the natural environment can be divided into three general categories, which may be termed decaying, cyclic and haphazardly fluctuating. Some systems always belong to one broad category while others change from one to another over relatively short periods of time.

DECAYING SYSTEMS

These consume their own substance, which may be energy or matter, or both. A good example is the decay of river-flow in dry weather, when the flow decreases each day but the rate at which the flow decreases also decreases with time and is proportional to the available water stored in the rocks. In this case the river-flow approximates to a negative-exponential decay curve, an example which is shown in Figure 1.2. Under such conditions the amount of substance

stored in the system will decrease to one-half of its original value in a given constant time interval. For example, the amount of available water stored in the rocks may be halved after every period of 10 days.

CYCLIC SYSTEMS

The input of short-wave radiation follows diurnal and annual cycles, and these are imposed on many natural systems. Heat balances of land surfaces are largely controlled by the input of solar energy, and therefore show both diurnal and annual cycles. Air temperatures reflect the state of the heat balance of the surface and therefore also show marked diurnal and annual cycles. The annual variations in solar radiation are least at the equator and increase towards the poles, and this leads to an increase in the annual variation of temperature with higher latitude. The variations in many cyclic systems when observed over a period of time appear to approximate to a mathematical curve known as a sine curve, which may be obtained by plotting the sine of an angle against the angle itself.

HAPHAZARDLY FLUCTUATING SYSTEMS

Some systems fluctuate in a random and irregular manner, changes occurring at unpredictable times and by unpredictable amounts. Turbulence in fluids or the occurrence of earthquakes are good examples, since neither can be exactly predicted. On small space and time-scales most systems exhibit some degree of unpredictability.

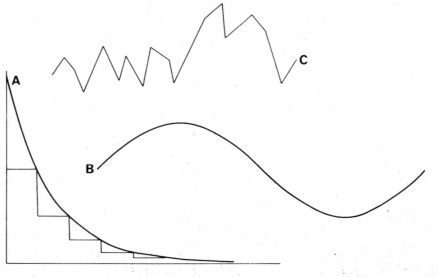

Figure 1.2 Decaying, Cyclic and Haphazardly Fluctuating Systems.

Curve A. An example of a negative—exponential decay curve. If the curve represents the amount of substance in a store, then it will decrease to one half of its original value in a given constant time interval.

Curve B. Sine curve.

Curve C. A haphazardly fluctuating curve.

The Ecosystem

It now remains to consider the place of the biological world in the natural environment, and this is best done through the concept of the ecosystem. The term 'ecosystem' was formally proposed by the plant ecologist Tansley (1935) as a general term for both the biome (the whole complex of organisms—both animals and plants—naturally living together as a sociological unit) and its habitat. All the parts of such an ecosystem—organic and inorganic, biome and habitat—may be regarded as interacting factors, which, in a mature ecosystem, are in approximate equilibrium.

The ecosystem concept has four main properties which, according to Stoddart (1965) make it a useful idea for use in physical geography.

(i) It is monistic, that is to say, it brings together environment, man, and the plant and animal worlds within a single framework, within which the interactions between the various components can be analysed. This is important because earlier in this chapter, comment was made that physical geography is largely concerned with the study of the natural environment as it relates to man.

(ii) Ecosystems are structured in an orderly, rational and comprehensible way. Once the structures are recognized, they may be investigated and measured by the normal techniques of the field sciences.

(iii) Ecosystems function—that is to say, they involve continuous through-puts of matter and energy which again can be measured.

(iv) In system terms, the ecosystem is an open system tending towards a steady state.

Most ecosystems involving man are extremely complex and therefore difficult to investigate and describe. It is more usual to investigate the very limited systems found on isolated islands or involving primitive human groups. Many studies concentrate on changes in the ecosystem caused by human interference, or by the introduction of new animals.

The great contribution made by the ecosystem idea is the constant stress placed on the fact that interactions and flows extend throughout the natural environment. Changes in one part of the ecosystem can have consequences which extend throughout the local ecosystem and can even have an influence at some distance. A good example is the use of certain insecticides on farm crops which can indirectly lead to a drastic reduction in the bird population.

Tropical Ecosystems

Before starting to explore ecosystems in detail, it is necessary to consider what are the important aspects of tropical ecosystems and what in particular makes them distinct from those of the temperate latitudes.

There is no one type of tropical ecosystem any more than there is just one type of temperate ecosystem. Instead, a great variety of ecosystems are found in both of the broad climatic zones, and some in each of the zones have properties which really belong to the other zone. Nevertheless, it is possible to distinguish a series of characteristics which separate the majority of ecosystems found in the tropics from those found in middle or high latitudes.

Probably the most important characteristic of tropical ecosystems is that they have a high radiant energy input. Solar radiation values are high throughout the year within the tropics and usually show only slight seasonal variations, and similar comments also apply to the net radiation (total incoming − total outgoing radiation). Actual solar and net radiation values at some middle latitude localities during summer may be higher than those observed in the tropics, but these high values are not maintained throughout the year. Indeed, the same middle latitude sites may have large net radiation deficits in winter. Tropical ecosystems are therefore characterized by high radiation fluxes throughout the year.

Net radiation can be used to heat the soil and atmosphere causing high temperatures, or to evaporate water from the oceans and vegetation. Usually net radiation is used firstly to evaporate water and only secondly to warm the atmosphere or soil. Thus very high temperatures can only exist over dry surfaces, since over wet surfaces the temperature is kept below about 32°C by the process of evapotranspiration. Temperatures will always be below about 32°C over large tropical oceans, which will have warm but not extremely hot climates, while over tropical land masses the amount of available moisture and therefore the general temperature levels are controlled by the rainfall.

Two extreme tropical climatic types can be recognized—firstly a rainless desert and secondly a type with a high rainfall spread evenly throughout the year. In the rainless desert there can be no evaporation and therefore the whole of the net radiation is available for heating the atmosphere and the soil.

Under these conditions midday temperatures may reach 50°C, but both diurnal and seasonal ranges are large. This type of climate is found in the sub-tropical deserts, which are often discussed under the general heading of the arid tropics.

In areas of the second climatic type with plentiful rainfall, much of the net radiation is used to evaporate water and therefore temperatures remain both relatively moderate and constant. Because of the abundant rainfall such areas usually have a natural vegetation of forest and are referred to as the humid tropics. Much of South-East Asia belongs to the humid tropics and it has a mean annual lowland temperature of around 26°C with both diurnal and seasonal variations of just a few degrees.

Within the lowland tropics both temperatures and net radiation are always adequate for plant growth, and are never limiting factors as in temperate latitudes. Usually the limiting climatic factor for plant growth and agricultural production is rainfall, and therefore rainfall seasons are of great importance.

Some tropical regions have a plentiful rainfall throughout the year, and thus are continuously humid, but others have very marked wet and dry seasons resulting in an alternation between humid tropical and arid tropical conditions. In these cases the dry season is often also the hot season because temperature levels are not moderated by evapotranspiration. When the rainy season starts temperature often falls, both because of the increase in cloud and also the energy loss due to evapotranspiration. A good example is India where over large areas the highest temperatures are observed just before the start of the wet season in June. Within middle and high latitudes, climatic seasons are not caused by rainfall variations but by the large seasonal changes in net radiation. Indeed, in the interiors of many temperate continents the highest rainfall is often associated with the warm season when convective activity and evapotranspiration are at a maximum.

A further aspect of tropical climates which may be of importance in the development of landforms is the intensity of rainfall. As a general rule rainfall intensities are greater in warm humid climates than in cold dry ones, so some of the world's highest rainfall intensi-ties and totals are observed in the tropics. Even the relatively dry areas can suffer from extremely intense rainstorms which cause extensive surface run-off.

Many chemical and biological reactions proceed at rates which depend on the temperature and the amount of solar radiation. Some chemical reactions double in speed with a 10°C rise in temperature. Thus the continuously high temperatures within the tropics allow chemical and biological changes to proceed at much greater rates than those found in temperate latitudes. This is further accentuated by the great age of many tropical ecosystems. The last glaciation completely destroyed the vegetation communities and soils over large areas of the temperate latitudes and therefore the age of many temperate ecosystems is considerably less than 10 000 years. Within the tropics there were changes associated with the temperate latitude glaciations, but they were not of a widespread destructive nature, and so in the tropical core regions the ecosystems may have only undergone slight changes during long periods of time and may even have survived basically unmodified from the late Tertiary. Therefore both the greater speed of reactions and the greater age of tropical ecosystems have allowed them to develop structural features which are not normally found in temperate latitudes. A good example is the presence of laterite in tropical soils.

FURTHER READING

Chorley, R.J. and Kennedy, B.A..(1971). *Physical Geography, A Systems Approach* (Prentice-Hall International Inc., London).

Eyre, S.R. (1964). 'Determinism and the Ecological Approach to Geography'. *Geography*, 49, pp. 369-76.

Stoddart, D.R. (1965). 'Geography and the Ecological Approach. The Ecosystem as a Geographic Principle and Method'. *Geography*, 50, pp. 242-51.

Tansley, A.G. (1935). 'The Use and Abuse of Vegetational Concepts and Terms'. *Ecology*, 16, pp. 284-307.

Freeman, W.H. (1970). 'The Biosphere'. *Scientific American*, San Francisco.

2   The Structure of the Earth

EARTH, which is a planet in orbit around the sun, consists of a massive solid core surrounded by a shallow atmosphere and a discontinuous ocean. Hypotheses of the earth's origin tend to favour the process of accretion of the earth and other planets through the coming together of cold clouds of dispersed gases and dusts under gravitational attraction. It is considered that at the time when accretion was largely complete, the earth's interior was not molten, though local melting may have taken place because of the energy released on impact. Some of the naturally-occurring elements are radioactive, that is to say, their atoms spontaneously disintegrate forming new stable elements, a process which is accompanied by the emission of various types of radiation, for example, uranium-235 decays to produce the stable elements of lead-207 and helium. The time taken for the activity of a given quantity of a radioactive element to decrease to one-half of its original value, or for half of the atoms originally present to disintegrate, is known as the half-life of that particular element. Typical half-lives of radioactive elements range from a fraction of a second to many millions of years, that of uranium-238 being 4·5 thousand million years (4·5 x 10^9 years). Many of the natural radioactive elements produce some heat when they decay, and this is the source of the warmth for the earth's interior. The heat flow at the present day from the earth's interior to the surface amounts to about 50 cal/cm^2/yr, which is extremely small compared with heat received at the surface from the sun.

Radioactivity was at a maximum at the time of the earth's formation about 4·5 x 10^9 years ago, and has decreased as decay reduced the supply of the radioactive elements. At first the radioactive elements were evenly distributed throughout the earth and they produced so much heat that the original materials melted. Differentiation within the molten mass produced the layering now observed, with iron concentrated in the core, surrounded by a less dense ultrabasic rock mantle, and a basaltic crust on the outside. The differentiation also concentrated the radioactive elements near the surface. With time the heat produced by radioactivity decreased and the earth became stable. So today the heat produced in the core by radioactivity is negligible, while the rate of surfaceward flow of heat from the upper mantle and crust closely balances the rate of heat production. Thus the mantle remains for the most part at a temperature lower than its melting point, though melting can occur in a shallow layer.

Between depths of 60 to 200 km the mantle rocks are at a temperature which is very close to their melting point. They are therefore in a plastic condition and will deform and flow if a steady stress is applied for a period of time, but they act as rigid bodies to sudden intense stresses such as earthquake shocks. This plastic layer in the mantle is termed the asthenosphere, while the strong, rigid overlying zone of the upper mantle and crust is called the lithosphere.

Isostasy

Existence of a plastic layer in the mantle allows individual segments of the lithosphere to rise or sink in response to unequal stresses, and the term isostasy is used to describe the condition of flotation balance that controls the heights of continents and mountains, and also the depths of the ocean floors. Thus the depth of the crust is greatest under the highest mountain masses and least under the ocean floors. Mountains are buoyed up by their deep roots in the plastic layer of the mantle. Denudation of mountains produces changes in load and results in a very slow isostatic adjustment to the new conditions. Since the removal of mass from a mountain range causes the mountains to float isostatically upwards, the net lowering by denudation is considerably less than might be expected.

Denudation rates vary with the average elevation of the surface and with the nature of vegetation cover. Typical rates for a continental surface of moderate or low average elevation, covered by forest and under a humid climate, are 6 to 7 cm per 1 000 years. In a region of high mountains and plateaux under an arid climate with scant vegetation cover, the rate is about 15 cm per 1 000 years and may locally reach 150 cm per 1 000 years. Maximum rates of crustal uplift are probably about 750 cm per 1 000 years, so extensive mountain ranges can be produced in the presence of denudation.

The process of denudation may be illustrated by an example. Consider a mountain range which has been elevated to a mean altitude of 5 km before uplift ceased. At first the average rate of denudation may be about 100 cm per 1 000 years, but as the elevation of the mountains is decreased so also is the rate of denudation, so an assumption can be made that one-half of the available land mass is removed every 15 million years. Much of the loss of altitude will be made good by isostatic adjustment due to the unloading, so the initial rate of net lowering of the surface may only be about one-fifth of the denudation rate, or about 0·2 mm per year. After 15 million years, when one-half of the available mass has been removed, the elevation is reduced to 2·4 km and the net rate of lowering to about 9 cm per 1 000 years. Similarly, after a further 15 million years, the mean elevation becomes about 1 km and the net lowering rate about 5 cm per 1 000 years, at which stage three-quarters of the original available mass has been removed. The decrease in elevation follows therefore a negative exponential decay curve as shown in Figure 1.2, and so after 60 million years the average elevation is about 0·3 km and the net lowering rate about 1 cm per 1 000 years.

Material eroded from the continents is deposited in the oceans where its accumulation causes the ocean floor to sink by the process of isostatic adjustment.

Peneplains

After an initial uplift a land mass becomes very rugged with many steep sided slopes and valleys, but with the process of denudation, the landscape is not only lowered but the overall ruggedness of the landscape decreases and eventually it becomes very subdued in aspect. From the nature of the exponential-decay curve, which describes the average lowering of a landscape, it is evident that zero elevation can never be reached, instead the landscape will become very subdued but denudation will continue at a very slow rate. W.M. Davis called such a landscape a peneplain from the two words—penultimate and plain. In the denudation example described above, the land surface can be considered to represent a peneplain when the average elevation is about 0·3 km—a stage which is reached after about 60 million years. Since mountain building is a comparatively rare geological activity, there has been ample time during the earth's history for extensive peneplains to form.

Davis recognized three distinct stages in the geomorphological cycle in a humid climate and these stages may be termed youth, maturity and old age. Youth corresponds to the initial period of uplift when there is rapid erosion. The valleys are strongly V-shaped and lithological variations in the rocks cause waterfalls and rapids. In maturity, the drainage system becomes more integrated, with the waterfalls and rapids evident in youth disappearing. In a fully matured river it is possible to recognize three tracts. At its head is the mountain tract which is either a gorge or a V-shaped valley with steep sides. In the middle stages of the river is the valley tract where the gorge has opened up and the slopes are gentler and the valley is wider. The middle tract merges into the plain tract, where at the bottom of the valley there is now a flood plain, which is flooded whenever there is a high discharge of water. In old age, valleys become very broad and most of the relief has disappeared, leading to a gently undulating plain which Davis called a peneplain.

Davis assumed that the initial uplift provided a given amount of potential energy, and that as denudation proceeded the energy available in the landscape decreased until at the stage of peneplanation the amount of free energy available reached a minimum. The landscape is considered as an isolated system and the peneplain corresponds to the condition of maximum entropy when the available potential energy approaches zero. The assumption here is that material moves from the uplands towards the lowlands by some form of mass movement under the action of gravity. This is an extremely slow process, as is shown by a study of the moon which lacks both running water and an atmosphere, and where mountains rising up to 11 km above the surrounding plains have existed for a great age (4×10^9 years). Both the surface of the moon and that of Mars, which also at present appears to lack running water, are different from that of the earth because they are extensively and heavily cratered. Older features on the moon appear to be slightly more subdued than newer ones but there is little evidence of the large-scale decay of features after their initial formation. Mass may be transported over the earth's surface by a variety of means, but the major transporting agent is running water. Indeed,

much of the earth's surface has been shaped by running water, hence the difference in landforms from that of the lunar surface. The water moving over the earth's surface originates as precipitation, and the flow of water from the uplands to the sea carries material in solution and in suspension. Thus denudation of the continents is largely an indirect result of the flow of water from the continental interiors to the oceans, and it is also part of an open system.

The importance of running water in shaping the surface of the earth can be illustrated by comparing it with the surfaces of the moon and Mars. The moon has no atmosphere or water and the original surface

1. View of the lunar surface from Apollo 16 while in orbit around the moon between 94 and 120 km.

2. View of small crater in lunar surface taken by Apollo 11 astronaut during a moonwalk.

still largely exists. The two lunar photographs (1 and 2) show a cratered surface which is unlike anything observed on earth, even in desert regions. The photograph of Mars (3) shows a more subdued landscape than that of the moon, but it still contains craters. A sinuous valley some 400 km long and up to 6 km wide is visible. This feature resembles the outline of a meandering river and may have been formed by flowing lava or water. Mars has a very thin atmosphere and though dry now could have been slightly wet in the past. Again the surface is unlike that observed on earth. Thus our landscape is very much the product of erosion, particularly by water and the atmosphere.

Open systems are observed to increase the complexity of their internal organization with time and they can also achieve a steady-state condition. If the landscape is viewed as part of an open system, it is likely that, though there will be a removal of mass with time, there will also be a tendency for a growing complexity within the landscape and for a relative permanence in some of the landscape forms. Thus the complexity and variety of soil types and structures will increase with the increasing age of the landscape. Also, since gradients have to exist to allow water to drain towards the sea, even ancient landscapes will contain slopes and probably extremely complex landforms. Therefore, peneplains, though areas of low

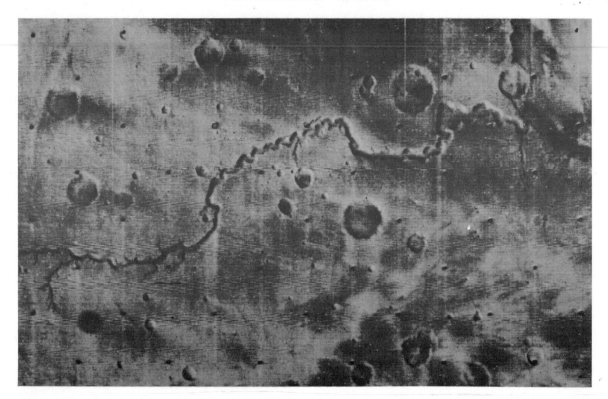

3. Television picture of Martian surface from Mariner 9. Photographs 1, 2 and 3 have been provided by the National Space Science Data Center through the World Data Center A for Rockets and Satellites.

relief, can contain complex landforms and soil structures and are not areas without interest. Structural movements of the earth's crust can further add to the complexity of forms found on ancient land surfaces. Much of the African and Australian continents consist of ancient surfaces which have been exposed to weathering for many millions of years. Africa, in particular, has been exposed to various types of tropical weathering since the beginning of the Tertiary, and is therefore a good example of a landscape of great age.

Rocks

The solid crust of the earth is composed of rocks and sediments in great variety, but extensive observations suggest that they can be grouped into three major categories, igneous, sedimentary and metamorphic, reflecting the manner of formation.

IGNEOUS ROCKS

Igneous rocks were caused by the cooling and solidification of a hot liquid that resulted from the melting of rock materials deep within the earth. Crystals of various minerals grow in the liquid as it cools, producing an interlocking arrangement of grains which eventually form an extremely hard rock. Since the exact size and type of mineral crystal formed depends on the rate of cooling and the temperature, a variety of igneous rocks can be produced from the same original liquid. There is no widespread extensive source of liquid rock, or magma as it is termed, near the surface, but instead it seems to form locally near zones of crustal activity, such as the plate junctions which are described later in this chapter. When magma wells up from a local source it may penetrate existing rocks to form layers or blocks of igneous rock or it may spread over the surface from a volcano as lava.

SEDIMENTARY ROCKS

Sedimentary rocks are composed of debris or pre-existing rocks that have been broken down chemically or physically by processes acting at the earth's surface. The material is normally transported by some agent, such as running water, from the site of the original rock to another place where the sedimentary rocks accumulate. Most sedimentary rocks accumulate in the sea as a series of horizontal beds or strata, but they can also form on land. The original deposited material usually consists of loose unconsolidated fragments, but with time it is often changed by percolating chemical solutions and by internal pressure into relatively hard rocks such as sandstones and shales. Sediments consisting of fragments of pre-existing rocks are known as clastic sediments. Sedimentary rocks can also be formed by chemical precipitation, which gives rise to some forms of limestone.

METAMORPHIC ROCKS

Metamorphic rocks are formed by the alteration of previously existing sedimentary and igneous rocks, usually deep below the earth's surface. In regions of intense geological activity, rocks are forced deep below the surface of the earth, where they are modified by the increased temperature and pressure. In this way, limestone can be changed to marble and sandstone to quartzite.

THE ROCK CYCLE

The various rock types can be connected together in a rock cycle, since igneous rocks become trans-

TABLE 2.1
GEOLOGICAL TIME-SCALE

Era	Period	Millions of years ago
Cenozoic	Quaternary	0–3
	Tertiary	3–70
Mesozoic	Cretaceous	70–135
	Jurassic	135–180
	Triassic	180–225
Palaeozoic	Permian	225–270
	Carboniferous	270–350
	Devonian	350–400
	Silurian	400–440
	Ordovician	440–500
	Cambrian	500–600
Precambrian		600– ?

formed to sedimentary rocks by weathering, erosion and deposition. These sedimentary rocks in turn can be metamorphosed, and the metamorphic rocks may melt to form a magma which will cool and crystallize to form igneous rock. The cycle may be short-circuited because sedimentary rocks are eroded to form new sedimentary rocks, and igneous rocks are changed directly to metamorphic rocks.

GEOLOGICAL TIME-SCALES

Rocks can also be classified according to the period of geological time during which they were formed. The geological record is broken into a series of geological periods based on the particular fossil assemblages of flora and fauna which are characteristic of that time. A list of the periods, which is rather arbitrarily drawn up, is contained in Table 2.1.

Continental Drift and Plate Tectonics

The lithosphere, or rigid outer shell of the earth, about 100 km thick, rests on a weaker layer known as the asthenosphere. Because of movements in the asthenosphere, the outer shell of the earth has been fractured into a number of separate blocks or plates, which move relative to each other. The boundaries between plates are zones of weakness and are marked by geological activity in the form of earthquakes and volcanoes. Earthquakes do not occur uniformly over the surface of the earth but are restricted to plate boundaries, while the rocks and sediments of the plates themselves show little evidence that they are currently being deformed. Continents may be visualized as being rafted over the surface of the earth as plateaux of granite-like rock resting on the even larger and thicker plates.

A general mechanism for plate movement and therefore continental drift is shown in Figure 2.1. The first stage of the process is when a new convection current develops in the asthenosphere and causes a rift to develop in what was a continuous plate. As the convection currents continue the rift grows and the plates with their associated continents move apart. The surface of the developing rift is below that of the general level of the continents so it soon becomes covered by water and forms a new ocean. As the rift continues to develop, molten rock from the asthenosphere flows into it and spreads laterally to form new ocean floor. So as the plates move apart the

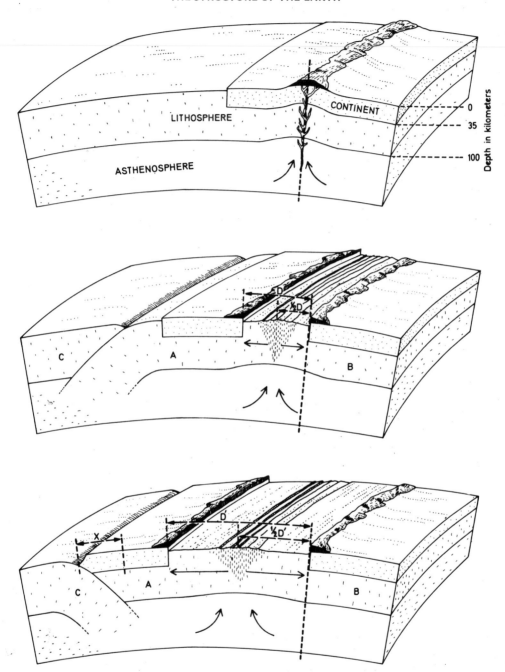

Figure 2.1 Mechanism of Continental Drift.

Upper. A continent is shown on a continuous plate (Lithosphere). This plate is split by motions in the Asthenosphere producing volcanic activity on the continental surface.

Middle. The continuous plate in the last diagram has now split into two separate plates A and B, and a new ocean is created between the plates. Since the floor of the ocean is spreading outward from a central mid-ocean ridge, the age of the ocean floor increases towards the continental margins.

Plate A is shown underriding plate C and being consumed in a subduction zone creating an oceanic trench and volcanic activity along the edge of plate C.

Lower. The continent on plate A encounters the trench and overrides it for some distance X and eventually reverses its direction from west-dipping to east-dipping. The ocean continues to grow and its width is now D′ as compared with D in the diagram above.

(After Dietz and Holden, 1970)

space created is filled by the formation of new material. The rock recently injected from below into the centre of the rift is slightly warmer than the old crust at comparable depth and the resulting lower density leads to a buoyant uplift and so produces a ridge along the centre of the rift. As the rock is pushed away from the ridge by the injection of fresh rock, the displaced rock cools, its density increases and it slowly subsides.

Good examples of rift oceans are the Atlantic and Indian Oceans which were formed when the continents split apart. Rates of sea-floor spreading measured in the Atlantic are remarkably rapid by geological standards, ranging from 1 to 3 cm/year, which is considerably faster than mountains are elevated. Similar rates of sea-floor spreading also apply to the Indian Ocean. It follows from this explanation of ocean formation that the central ridges of rift oceans should be geologically very young, and that the floor of the ocean should become progressively older with increasing distance from the central ridge. This is observed in the Atlantic, where the age of the igneous rocks of mid-ocean islands gradually increases with distance from the mid-ocean ridge.

Since new material and new surface area are being added to the plates in the rift oceans, it follows that the plates should lose material and surface area elsewhere, and this occurs where plates converge. Here one plate will underride an adjacent plate and dive down into the earth's mantle along subduction zones. If the rate of convergence is rapid, deep ocean trenches are formed by the downward movement of the subsiding plate. These trenches are typically about 10 km deep, 60 km wide and 1 500 km long, and occur near the continental margins. Increasing pressure and temperature with depth drive off the more volatile components which are exuded at the surface as volcanic lavas. Thus an oceanic deep may also be associated with extremely active volcanoes and earthquake activity. If rates of plate convergence are low, the underriding plate may not form an ocean deep but result instead in mountain building. The depth to which a surface plate can underride an adjacent plate is limited and therefore this mode of convergence normally ceases after about 1 800 km of ocean floor have been lost. At this stage the moving plate overrides the trench for some distance and eventually forces the dip of the trench to reverse with the motionless plate now being forced below the moving plate.

Most ocean trenches are found around the shores of the Pacific where they are associated with nearby volcanic mountains and with extensive earthquake zones. Deep ocean trenches are unlikely in rift oceans such as the Atlantic.

The above ideas, which are relatively new, on the nature of the surface of the earth, may be summarized as follows:

(i) The surface of the earth is broken into large, rigid plates, most of which have been moving for at least 100 million years.

(ii) Moving plates are created at the trailing edges and simultaneously destroyed at the leading edges.

(iii) These plates are created along ridges which traverse many of the major oceans.

(iv) The type of deformation at the leading edges is a function of the rate at which they come together. If it is slow, the plates buckle and form high-fold mountains. If it is fast, one of the leading edges plunges into the mantle to form a zone of earthquakes and volcanoes accompanied by a deep trench and usually an island arc.

(v) Most activity is around the edges of plates, very little happening within the plates. Volcanism within plates is rare, but often intense where it occurs, as in the Hawaiian Islands.

The Evolution of the Continents and Oceans

Although various research workers have produced reconstructions showing the evolution of the continents, the one described here is based on that of Dietz and Holden (1970). Later workers will change details, but the main outline of the evolution of the continents now seems clear.

Two hundred million years ago (Figure 2.2), the continents were grouped together in one universal land mass which may be termed Pangaea. This large land mass may have been assembled from several smaller masses at an earlier date, but evidence for this is scanty. In Figure 2.2, Panthalassa formed the ancestral Pacific Ocean, while the ancestral Mediterranean (Tethys Sea) formed a large bay separating Africa from Eurasia. India was not joined to Eurasia at this stage, but together with Australia, bordered the Antarctic. Sinus Australis, a southern bay off the Tethys Sea, separated India from Australia.

Two extensive rifts (Figure 2.3) were initiated in Pangaea about 200 million years ago. The northern

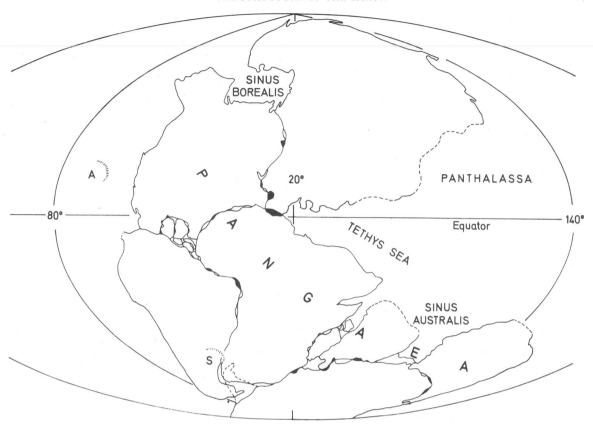

Figure 2.2 Continental distribution during Mid-Triassic.
During the Mid-Triassic period (200 million years ago) the universal land mass Pangaea may have appeared as shown. Panthalassa is the ancestral Pacific Ocean and the Tethys Sea the ancestral Mediterranean. India and Australia are shown joined to Antarctica but separated from each other by the gulf Sinus Australis. The hatch crescents serve as modern geographic reference points; they represent the Antilles arc in the West Indies and Scotia arc in the extreme South Atlantic. (After Dietz and Holden, 1970)

rift split Pangaea from east to west along a line slightly to the north of the equator and created Laurasia (composed of North America and Eurasia) and Gondwana. The southern rift split South America and Africa as a single land mass away from the remainder of Gondwana, consisting of Antarctica, Australia and India. A little later a secondary rift developed separating India from Antarctica.

During the period up to 135 million years ago (end of Jurassic) the rifts increased in size forming the Atlantic and Indian Oceans. About this time an incipient rift began separating South America from Africa, starting in the south and working north. The Tethys Sea formed not only a zone of crustal subduction, or trench, towards which India drifted, but also a zone of shear along which Eurasia slid westward with respect to Africa.

About 65 million years ago (end of Cretaceous,

Figure 2.4) the rupture of South America and Africa was complete and both the North and South Atlantic were in existence. Africa had drifted about 10° northward and rotated counter-clockwise as the Eurasian plate rotated slowly clockwise, resulting in the near closure of the eastern end of the Tethys Sea. India continued its progress north towards Eurasia, while a new rift carved Madagascar away from Africa, but Australia remained attached to Antarctica.

During the Tertiary, that is from about 65 million years ago to the present, the continents drifted to the positions that are observed today (Figure 2.5). India completed its journey northward by colliding with Asia, the northern margin of the Indian plate being subducted below the Asiatic plate, creating the Himalayas. During India's passage to the north in the early Tertiary its western margin crossed a fixed source of magma rising from the earth's upper mantle near the

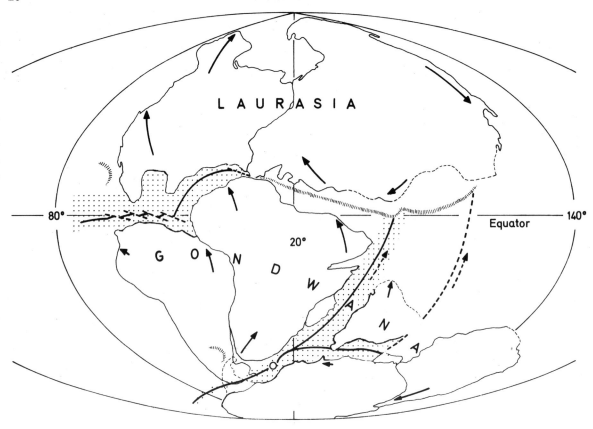

Figure 2.3 Continental distribution at the end of the Triassic. At the end of the Triassic period (180 million years ago) the northern group of continents (Laurasia) had split away from the southern group (Gondwana). India is shown moving away from Antarctica.
(After Dietz and Holden, 1970)

```
::::::::       New ocean floor
←———         Vector motions of continents
– – –          Zones of slippage along plate boundaries
·····››››·–‹    Tethyan trench
  o            Thermal centre
```

equator, and as a result molten rock erupted through the crust and poured onto the Indian subcontinent, laying down the basalt of the Deccan plateau. Even after India had moved north, magma continued to stream out over the ocean floor, producing the Chagos-Laccadive ridge, which became covered with coral as it subsided into the Indian Ocean. Also during the Tertiary a branch of the Indian Ocean rift split Arabia away from Africa, creating the Gulf of Aden and the Red Sea, with a spur of this rift meandering west and south into Africa.

Present Structure of the Continents and Oceans

Crustal movements over the last 200 million years

have led to the development of ten major plates plus numerous additional sub-plates. The major plates (Figure 2.6) may be listed as follows:

(a) Pacific plate which consists of most of the Pacific Ocean and includes small fragments of the west coast of North America.

(b) North American plate.

(c) South American plate.

(d) Eurasian plate which is almost entirely continental, consisting of all of Europe and most of Asia, including much of Indonesia and the Philippines.

(e) African plate.

(f) East African sub-plate which consists of Africa east of the rift valley.

(g) Madagascar plate.

(h) Indian plate which contains the portions of

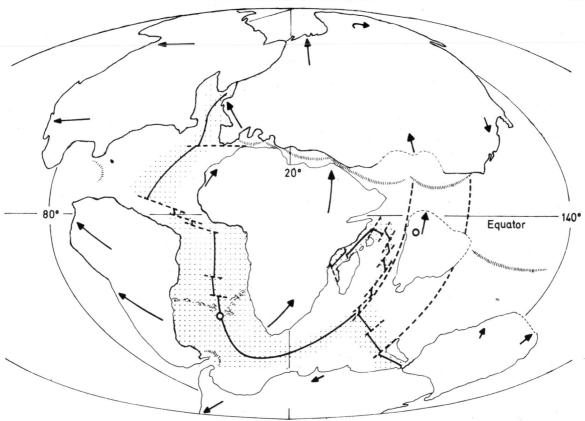

Figure 2.4 Continental distribution at the end of the Cretaceous. At the end of the Cretaceous period (65 million years ago) the Atlantic is clearly visible as a major ocean, created by the westward drift of the North American and South American plates. An extensive north-south trench (not shown) must have existed in the Pacific to absorb the westward drift of these two plates. The Indian plate is passing over a thermal centre which pours out basalt to form the Deccan plateau and later creates the Chagos-Laccadive ridge. A similar thermal centre exists in the South Atlantic and forms the Walvis and Rio Grande ridges.
(After Dietz and Holden, 1970)

Asia to the south of the Himalayas as far as Burma in the east, and also portions of the central Indian Ocean.

(i) Australian plate which contains Australia, New Guinea and New Zealand.

(j) Antarctic plate which is composed of the continent of Antarctica and the Southern Ocean.

Within southern and eastern Asia there are several plate junctions (see Figures 2.5 and 2.6) which are important elements of the relief. The junction between the Australian and Asian plates runs between New Guinea and Celebes, then to the south-west of the islands of Sunda, Java and Sumatra, and into the Bay of Bengal where it forms the boundary between the Indian and Asian plates. The Australian plate is moving slowly north and underrides the Asian plate along the junction to the south-west of Indonesia

forming the deep Sunda trench, and causing mountain building and volcanic activity in Sumatra, Java and their associated islands.

The Eurasian plate joins the Pacific plate along a series of deep trenches—the Philippine, Marianas, Bonin, Japan and Kuril trenches. These trenches are particularly deep, mostly over 10 km, and are associated with frequent earthquakes and volcanic activity. The boundary between the Australian and Pacific plates is marked by a series of trenches running north from New Zealand to the east of Tonga and the west to New Guinea. In particular the Kermadec and Tonga trenches are particularly deep and are associated with both earthquakes and volcanic activity.

Activity along plate junctions explains the distribution of both volcanoes and earthquakes in southern and eastern Asia. Away from the junctions the plates

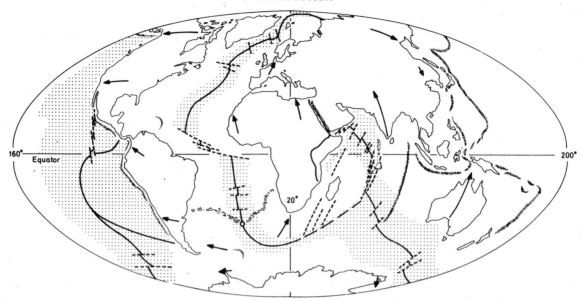

Figure 2.5 Continental distribution today. Nearly half of the ocean floor (shown stippled) was created in the Cenozoic. Major ocean trenches and zones of subduction are shown by hatched lines. Mid-ocean ridges indicated by thick lines. (After Dietz and Holden, 1970)

are stable and undisturbed, so active volcanoes and earthquakes are very rare in Borneo and Peninsular Malaysia which form part of the stable Eurasian plate, and in Australia.

Study of the development of the continents indicates that while the Atlantic and Indian Oceans are rift oceans, the Pacific is not a rift ocean, but the ancestral ocean which is becoming smaller as the new

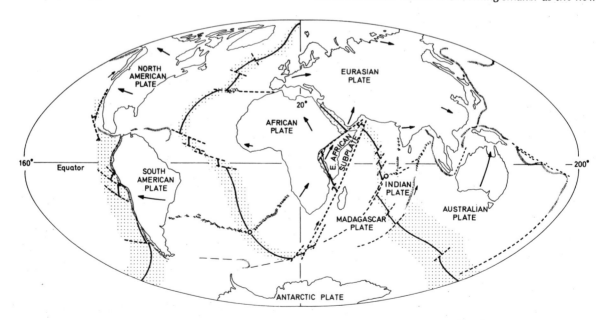

Figure 2.6 Continental distribution 50 million years from now. Continental distributions produced by extrapolating present-day plate movements.
(After Dietz and Holden, 1970)

ocean basins grow. Mid-ocean ridges are associated with the rift oceans and these ridges are the source of some volcanic and earthquake activity. The Mid-Atlantic ridge (Figure 2.5) traverses the entire Atlantic Ocean and separates the deep waters into eastern and western basins. On leaving the South Atlantic it turns half-way between Africa and Antarctica, splitting in the Indian Ocean into two branches, one of which heads between India and Madagascar into the Red Sea and the other between Australia and Antarctica.

Sea-level along the margins of the continents is not fixed and has actually varied very widely during the recent geological past. A fall in sea-level of about 200 m would expose a wide continental shelf, the Sunda shelf, connecting the Malay Peninsula to Borneo and a similar shelf, the Sahul shelf connecting Australia to New Guinea. These shelves were probably partly exposed during the low sea-levels associated with the last ice age. There is considerable evidence that both the creation of new oceanic crust by sea-floor spreading and volcanism have been unusually intense during the last ten million years. Volcanoes are sources of carbon dioxide, fine ash and water-vapour, the latter increasing the volume and therefore the depth of the oceans. Since the actual volume of the ocean basins has also increased, probably by an amount greater than the volume of new volcanic water, there has been a long-term lowering of sea-levels during the late Cenozoic.

The Structure of South-East Asia

South-East Asia, which belongs to the Eurasian plate (Figure 2.7), may be regarded as having been progressively consolidated from the north-west towards the south-east. So while much of the mainland consists of pre-Tertiary formations, at least 70 per cent of the surface rocks of Indonesia are either Tertiary or Quaternary in age, and Java is almost completely Tertiary or later in age. The oldest surface rocks in the area are a partially exposed fragment of pre-Cambrian folding which occupies much of Cambodia and southern Vietnam, and it is around this ancient nucleus that the rest of South-East Asia was gradually built. Mountain building took place in Mesozoic times and structures of this age now predominate over the greater part of Thailand, lower Burma, and throughout the Malay Peninsula and

south-western Borneo. In Triassic times the great massif of the Sunda shelf came into existence, and it has remained generally stable, though its outlines have been modified by marine transgression, and at present it is covered by the shallow waters of the Malacca Straits and the South China Sea. Along the southern margin of the Sunda shelf are two series of folds which were formed during the Tertiary. The earlier, dating from about mid-Miocene times is responsible for the outer line of arcs, running through the western ranges of Burma, the Andamans and the lesser islands south-west of Sumatra, to the Moluccas and probably then into eastern Celebes. The inner and younger arc, which originated in the late Pliocene, runs roughly parallel to its predecessor, and may be traced through the medial hills of Burma, and the central axes of Sumatra, Java, and the main line of the Lesser Sundas. Scattered along this inner line are 200 or more volcanoes, of which 70 in Indonesia and a dozen in the Philippines have erupted during the last 150 years. Indeed, the main contrast between the older core and the peripheral Tertiary structures is the relatively subdued and erosional relief of the former and the bolder and more markedly tectonic relief of the latter.

The complex structures along the edge of the Sunda shelf are well displayed in Sumatra. In Sumatra the volcanic inner mountain arc is represented by the Barisan Range which lies a short distance from the west coast. These mountains show intense folding of sedimentary rocks and also block-faulting and volcanic activity. The islands west of Sumatra (Simeuluë, Nias, the Mentawai group, Enggano) belong to the partly submerged non-volcanic outer arc. They are composed of Tertiary deposits many thousands of metres thick, and their uplift began in the Quaternary. Seaward of these islands there is a deep ocean trough. Extreme south-eastern Sumatra, the islands of Bangka, Belitung and the Riouw-Lingga group, form part of the stable Sunda shelf, which differs markedly from the unstable remainder of Sumatra. These islands were reduced to an undulating peneplain and then submerged in the early Pleistocene.

Relief in Sumatra reflects several cycles of erosion in addition to crustal movements. Faulting and volcanic activity are the two most important phenomena affecting the geomorphology of the Barisan Range, and both are much in evidence today. Faulting produces block mountains, which often have rather flat

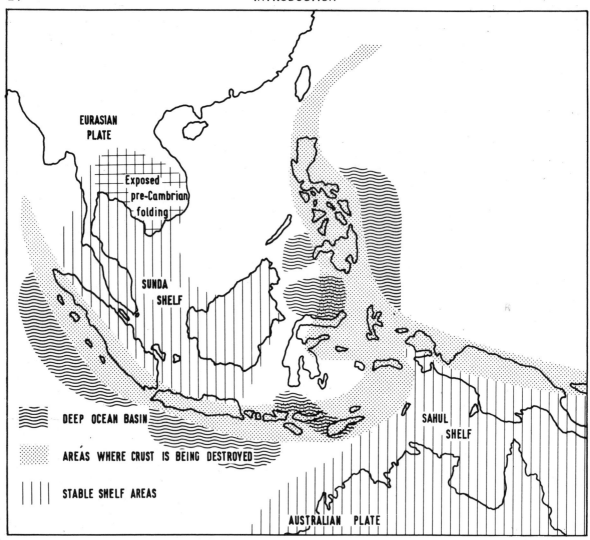

Figure 2.7 Graphical illustration of the structure of South-East Asia.

surfaces suggesting prolonged erosion prior to uplift. It also strongly influences the river systems, causing them to change direction and form terraces on the valley sides in areas of rapid uplift.

Many types of volcanic features are found in the Barisan Range. Basaltic lava tends to form rather flat volcanic cones or plateaux, an example being the Sukadana area of south Sumatra. Acid lava forms steep-sided cones and examples of these are found near Tandjung Kasang.

The Structure of the Malay Peninsula

The Malay Peninsula was given its present form and structure by the main geological events of the Mesozoic Era. At the beginning of this era, the region which extends from west Yunnan, through east Burma and west Thailand to the Malay Peninsula, had already experienced a long geosynclinal history, commencing as early as the Ordovician. Indeed, during the Permain much of South-East Asia was covered by a sea in which limestone-type rocks were deposited. The deposition was apparently centred in Thailand and extended into eastern Burma, southern China, Indo-China, and the north-west quadrant of the Malay Peninsula. The Permain conditions lasted into early Triassic time, as is shown by the massive limestones from this period in south Kelantan and north-west Pahang. The Triassic saw the final stages of the geosynclinal phase, for it was also the start of the culminating orogenic revolution of the region—the Thai-Malayan Orogency—which was accompanied by granite emplacement on a vast scale. This orogency seems to have had a revolutionary phase that lasted from the early Triassic to the early Cretaceous and to have had several pulses. Much of the granite has now been exposed by erosion and forms the Main, Kledang and Bintang ranges of the Malay Peninsula. When the orogency had finished, the Malay Peninsula had been transformed from a mobile to stable region, and henceforth formed a part of the Sunda Shield.

Throughout Cenozoic time, the southern part of the Malay Peninsula was largely emergent and relatively stable tectonically, activity being confined to uplift and tilting. Stauffer (see Gobbett and Hutchison, 1973) states that the known or suspected Cenozoic sediments include the following:

(i) A series of small basins containing Tertiary sedimentary deposits between the Straits of Malacca and the Main Range. The sediments consist of partly consolidated gravel and sand, soft shale, seams of low-grade coal, and sometimes limestones. All are low-lying, being barely above the level of the alluvial plains. So far, five such basins have been identified, and from north to south they are:

Bukit Arang-Betong;

Enggor;

Batu Arang;

Kepong;

Kluang-Niyor

(ii) Local basalt flows, near Kuantan in Pahang and at Segamat in Johore. They cover an area of about 150 km² around Kuantan, the highest point being at Bukit Tinggi (138 m). Both basalt flows are probably early Quaternary in age.

(iii) Widespread deposits of weathered gravel, sand, clay and peat, dating mainly from the early and middle Pleistocene and usually referred to as the Old Alluvium.

(iv) Pleistocene terrace deposits of weathered gravel and finer material along some of the major rivers.

(v) A locally conspicuous ash deposit, found mostly in the Perak River valley and western Pahang. Since the Malay Peninsula has no known Cenozoic source for ash of this type, it must have been blown by the wind from a nearby volcanic area. The catas-

trophic prehistoric eruption that produced the caldera of Lake Toba in northern Sumatra would appear to be the most likely source.

(vi) Extensive modern deposits, including only slightly weathered or unweathered gravel, sand and clay, inland, and multiple beach ridges and marine or lagoonal clays near the coasts, dating from the Holocene and late Pleistocene. They are usually referred to as the Young Alluvium.

HIGHLANDS

As already noted much of the Malay Peninsula is formed of deeply-dissected highland areas which have a long erosional history through successive uplifts since the Triassic. Granite rocks dominate the topography by forming all but one of the major mountain ranges with summits exceeding 2 000 m. The influence of faults in the granite is seen in rectangular drainage patterns observed in some areas.

A difference in morphology is observed between the granite rocks of the Johore-Singapore region and areas further north. In the south, granite regions are marked by undulating hill topography with elevations of a few hundreds of metres. Further north the topography of the granite is characterized by steep valley walls, numerous waterfalls and rapids, and small remnant upland plateaux, indeed all features indicating a late youthful stage in Davis's geomorphological cycle. This is the result of warping, mainly in the late Cretaceous and early Tertiary, during which the northern and eastern parts of the Malay Peninsula have been preferentially uplifted by as much as 1 000 m relative to the western and southern areas.

Cliff-bounded limestone hills rising up to 600 m above the surrounding country are prominent features of northern parts of the Malay Peninsula. The hills are usually honeycombed by caves and show a large range of karst features, which are discussed in Chapter 10. Two types of tropical limestone morphology can be recognized—tower or mogote karst (as observed in the Malay Peninsula) and sinoid karst which is composed of numerous low hillocks (as observed in central Java). Tower karst seems to develop on rocks with vertical joint planes undergoing extremely rapid vertical erosion associated with a deep water table.

OLD ALLUVIUM

Deposits of the Old Alluvium are found in many

coastal areas and also in inland valleys such as the Kinta Valley. They occur at elevations up to 70 to 75 m above sea-level and in some areas form low dissected hills. Fluvial sediments can be deposited at varying elevations above sea-level and need not be directly connected with the sea-level existing at the time of deposition. So it must not be automatically assumed that such deposits indicate a higher sea-level some time in the past. Nevertheless, widespread occurrences of the Old Alluvium near the present coast in Johore and Singapore, some of it possibly representing an estuarine mud-flat environment, do suggest the former existence of a coastal or piedmont plain graded to a base-level somewhat above the present sea-level. The deposits in the Kinta Valley and the Kuala Lumpur area do appear to be completely fluvial in origin, and probably formed in the higher tributaries of local river systems. The evidence available suggests that the Old Alluvium is Pre-Holocean, that is to say, mostly Pleistocene in age and that some of it may even date from the late Tertiary. There is considerable debate about sea-levels in the Pleistocene and this topic is considered again in Chapter 9.

MODERN DEPOSITS

There are unconsolidated deposits of sand and gravel with some peat and clay, known as the Young Alluvium, locally overlying the Old Alluvium and filling channels and depressions in it, as well as overlying bed rock at many other places. These deposits represent a thin capping of channel deposits from rivers and date from the last interglacial up to recent times.

Much of the coastline of the Malay Peninsula consists of a coastal plain of very low relief just a few metres above sea-level. The width of the plain varies from zero, where hills form the actual coast, to about 20 km along large parts of both the east and west coasts, and to a maximum of about 60 km around the mouth of the Perak River. The surface of the coastal plain is made up of beach, lagoon, swamp and river deposits which in many areas record a marked advance of the shore since the end of the last glacial period. Indeed, the result of the rise in sea-level during the last 15 000 years is apparent along the coast of the Malay Peninsula in the form of estuaries, offshore islands and wide swampy areas. Rates of coastal accretion are rapid in many areas, for instance the coastline south of Kuantan has advanced at a rate of about 16 m annually for the last 1 500 years and similar annual rates have been observed in the Dindings area of Perak.

Inland ridges composed of white sand with some shell fragments representing former beaches are a common feature of many parts of the coastal plain of the Malay Peninsula. Typically the ridges are long linear or curved features which tend to run parallel to the present coast and are separated by low-lying areas representing former lagoons or swamps. Near Kuantan the ridges are found up to 6 km inland and at elevations up to 11 m above existing sea-level. Most of the ridges appear to mark advances in the coastline since the end of the last ice age.

FURTHER READING

Dietz, R.S. and Holden, J.C. (1970). *Scientific American*, December, p. 30.

Gobbett, D.J. and Hutchison, C.S. (1973). *Geology of the Malay Peninsula* (Wiley-Interscience, New York).

Holmes, A. (1965). *Principles of Physical Geology* (Nelson).

Strahler, A.N. (1972). *Planet Earth: its Physical Systems through Geologic Time* (Harper and Row).

Wegener, A. (1924). *The Origin of Continents and Oceans* (Methuen).

PART B

Energy in the Tropical Environment

Cumulonimbus cloud. Latent heat is transformed into sensible heat and potential energy in these clouds.

3 The Nature of Energy

ENERGY may formally be defined as the capacity for doing work, and it may exist in a variety of forms including heat, radiation, potential energy, kinetic energy, chemical energy, and electric and magnetic energies. It is a property of matter capable of being transferred from one place to another, of changing the environment and is itself susceptible to change from one form to another. An example of energy changing the environment is provided by solar radiation falling on a field during the early morning and increasing both the temperature of the air and of the plants. Another example is the energy of high winds or floods which may change the natural environment in a far more spectacular and destructive manner. If nuclear reactions are excluded, it can be stated that energy is neither created nor destroyed, and from this it follows that all forms of energy are exactly convertible to all other forms of energy, though not all transformations are equally likely. It is therefore possible for any particular system to produce an exact energy account, in which the energy gained exactly equals the energy lost plus any change in storage of energy in the system. Since a continuous transformation of energy from one form into another takes place in the atmosphere and on the earth's surface, it is necessary to consider in some detail the various forms which energy can assume.

Heat

Heat is a form of energy and it defines in a general way the aggregate internal energy of motion of the atoms and molecules of a body. It may be taken as being equivalent to the specific heat of a body multiplied by its absolute temperature in degrees Kelvin and by its mass, where the specific heat of a substance is the heat required to raise the temperature of a unit mass by one degree. It is important to distin-

guish between temperature and heat, for temperature is a measure of the mean kinetic energy (speed) per molecule of the molecules in an object, while heat is a measure of the total kinetic energy of all the molecules of that object. As the temperature increases so does the mean kinetic energy of the molecules, and conversely it is possible to imagine a state when the molecules are at complete rest, a point on the temperature scale known as absolute zero. This has been found to be at 273·15 Celsius degrees below the melting point of ice ($0°C$), and the Kelvin temperature scale is measured from absolute zero in Celsius units making $0°C$ equivalent to $273·15°K$. The Kelvin temperature scale is used in basic physical equations which involve temperature and it has the practical advantage of avoiding negative values.

Temperature is the condition which determines the flow of heat from one substance to another, the direction being from high to low temperatures. So long as only one object is considered, its temperature changes represent proportional changes in heat content. The definition of heat content suggests that when a variety of masses and types of material are considered, the equivalence of heat and temperature disappears. Often a small hot object will contain considerably less heat than a large cool one, and even if both have the same mass and temperature their heat contents can differ because of differing specific heats.

The transfer of heat to or from a substance is effected by one or more of the processes of conduction, convection or radiation. The common effect of such a transfer is to alter either the temperature or the state of the substance or both. Thus, a heated body may acquire a higher temperature (sensible heat) or change to a higher state (liquid to gas, or solid to liquid) and therefore acquire latent or hidden heat. Conduction is the process of heat transfer through matter by molecular impact from regions of high temperature to regions of low temperature without the transfer of the matter itself. It is the process by which heat passes through solids but its effects in fluids (liquids and gases) are usually negligible in comparison with those of convection. In contrast, convection is a mode of heat transfer in a fluid, involving the movement of substantial volumes of the substance concerned. Conduction is the main method of heat transfer in the solid rocks and the soil, while the convection process frequently operates in the atmosphere and oceans.

Radiation

This is the transmission of energy by electro-magnetic waves, which may be propagated through a substance or through a vacuum at the speed of light. Electro-magnetic radiation is divided into various classes which differ only in wave-length; these are in order of increasing wave-length—gamma radiation, X-rays, ultraviolet radiation, visible radiation, infra-red radiation and radio waves. All objects which are not at the absolute zero of temperature give off radiant energy to the surrounding space, so the natural environment is full of radiation of various wave-lengths, the most important of which are in the visible and infra-red sections. Furthermore, since nearly all the available energy in the natural environment was originally gained as visible radiation from the sun, the study of radiation is obviously of great importance and is considered separately in greater detail in the next chapter.

Potential Energy

This is the energy possessed by a body by virtue of its position. It is measured by the amount of work required to bring the body from a standard position, where its potential energy is zero, to its present position. Thus a body at some distance above the ground has more gravitational potential energy than a body at ground level, and if released the potential energy will be converted into kinetic energy as the object accelerates towards the earth. Rivers are good examples of the conversion of potential energy into kinetic energy. Water-vapour in the atmosphere possesses some gravitational potential energy in respect of its altitude above sea-level, and this potential energy is converted into kinetic energy when it condenses into

rain which then falls towards the ground. If the rain reaches sea-level, then all the gravitational potential energy of the rain water will have been converted into kinetic energy, but in contrast, if it falls on an upland surface, some potential energy will still be available and this will appear as the energy of river-flow as the water moves towards the sea along stream channels. Viewed in this way, the energy for the erosive power of rivers comes originally from gravitational potential energy.

Kinetic Energy

This is the energy possessed by a body by virtue of its motion. It is a quantity of magnitude $\frac{1}{2}MV^2$, where M is the mass and V the velocity of the particle. Kinetic energy is continuously dissipated by the various resistances to motion, and is often converted into heat. The kinetic energy of rivers is dissipated by the resistance to water movement created by the uneven stream floor. Since this resistance is very large, rivers normally flow only very slowly, suggesting that kinetic energy is being destroyed almost as fast as it is created from the gravitational potential energy of the water. The atmosphere contains kinetic energy because of the winds, and this is dissipated mainly by friction at the ground surface. It is estimated that, in the absence of solar radiation which creates kinetic energy, dissipation of the atmosphere's kinetic energy by friction would be almost complete after six days. Kinetic energy is therefore, in the natural environment, one of the less stable and more short-lived forms of energy, and will soon be converted into other forms unless it is continually renewed.

Chemical Energy

This is the energy used or released in chemical reactions. Some chemical reactions, such as the process of burning or exploding, release large amounts of energy in the form of heat and light, whereas others absorb or release energy only very slowly. In the natural environment there are various chemical reactions, one of the more important being the process of photosynthesis in plant leaves. In photosynthesis the elements of two atmospheric gases (carbon dioxide and water-vapour) are combined with light energy captured by the chloroplasts of plant leaves to form plant materials and oxygen.

$$CO_2 + 2H_2O \xrightarrow[\text{green plant}]{\text{light}} (CH_2O) + O_2 + H_2O$$

Therefore plants normally require the energy contained in sunlight to grow. If plant materials are burnt, the energy originally absorbed over long periods of time is released in a short intense burst of heat and light. The amount of photosynthesis which takes place on earth may be estimated from the amount of carbon fixed from carbon dioxide each year. Estimates of this type show that 90 per cent of carbon is fixed by aquatic plants and the remainder by land plants, of which forests fix 7 per cent leaving only 3 per cent for all the managed and unmanaged fields on earth.

Electric and Magnetic Energies

These are of little importance in the natural environment at the earth's surface, but they are of great interest in the very high atmosphere on the edge of space. The most obvious manifestation of electric energy near the surface is provided by lightning in thunderstorms.

Energy Transformations

Continual transformations of energy from one form to another take place in the atmosphere and on the earth's surface. These energy transformations are largely responsible for creating the natural environment, because without energy the world would be completely dead, for there would be no movement or life. Some of the more important transformations are discussed below.

(a) Radiation⟶heat⟶radiation.

Solar radiation falling on a forest causes an increase in the heat content of the soil, the plants and the air, and this is observed as a rise in temperature. The forest also loses heat as infra-red radiation.

(b) Radiation⟶(sensible heat + latent heat).

This is a variation of the first case. If the radiation falls on water or a moist surface (plants, etc.), some of the radiant energy will be used in warming the surface, forming sensible heat, but some will be used in evaporating water-vapour and this will form latent heat. Sensible heat is heat that can be felt and detected by a change in temperature, whereas in con-

trast, latent heat is completely hidden and is only detected when a change of state of the substance concerned takes place.

(c) Latent heat⟶sensible heat⟶radiation.

This transformation occurs when water-vapour condenses into water droplets. The latent heat taken up at evaporation is released and appears as sensible heat which will eventually form radiation. This process is observed in clouds where the sensible heat released warms the surrounding atmosphere.

(d) Potential energy ⟶ kinetic energy ⟶ heat ⟶radiation.

This sequence is a very common one, since kinetic energy is usually dissipated by friction and is converted into heat. It was noted earlier when the example of river-flow was considered.

Now all the above transformations of energy can proceed in the reverse as well as the forward direction. Thus kinetic energy can become potential energy, sensible heat can become latent heat and so on, and indeed the only transformation which is not usually observed is the direct conversion of radiant energy into kinetic energy.

Energy in the Atmosphere

Energy transformations which occur in the natural environment are well illustrated by a study of the atmosphere where the various components of energy may be found. The energy content of one gramme of moist air may be written as follows:

Total energy content (Q) = Latent heat content (Lq) + sensible heat content (C_pT) + potential energy content (gz) + kinetic energy content $(V^2/2)$.

In the following discussion it is assumed that one gramme of moist air is under discussion, so mass will often be omitted. The first component of the total energy content is the latent heat term, which consists of the latent heat of vaporization (L) multiplied by the mass of water-vapour (q). Latent heat is formally defined as the quantity of heat absorbed or emitted, without change of temperature, during a change of state of unit mass of a material and for vaporization of water is equivalent to 575 cal per g. Since one calorie is the amount of heat required to raise the temperature of 1 g of water by 1°C, it is seen that a large amount of energy is locked as latent heat in water-vapour. Indeed, in moist tropical air the latent heat component can amount to about 15 per cent of the total energy content.

Sensible heat is the product of the specific heat (C_p) and the temperature (T) in degree K, and is contained in the second term. The atmosphere is continually losing sensible heat, in the form of radiation to space, at a rate of about $\frac{1}{4}$ per cent per day of the total energy content; this represents a cooling of about 1 to 2°C per day. Sensible heat is gained from the surface and by the release of latent heat from condensing water-vapour.

Potential energy exists in the unit parcel by virtue of its position above the earth's surface. It is the product of the height z above sea-level and the force of gravity g. If the parcel sinks slowly, the potential energy must decrease and reappear as another form of energy, which normally takes the form of sensible heat. In the atmosphere there is a very close relationship between potential energy and sensible heat, since as air parcels sink, their potential energy is converted into sensible heat and as they rise, their sensible heat is converted back into potential energy.

An adiabatic process is one in which heat does not enter nor leave the system, and so there are no radiative gains or losses, no heat flows, no mixing with the surrounding environment, and no changes in water content. Thus the total energy content of a unit air parcel will remain constant during an adiabatic change. Furthermore, if there is no condensation nor evaporation within the parcel, that is to say, if the latent heat component remains constant, then under adiabatic conditions the sum of the sensible heat and the potential energy will also remain constant, and it is convenient to consider these two quantities together under the heading of total potential energy. Since the total potential energy is constant in an adiabatic process, the temperature of a rising air parcel must decrease since this is the only way in which the sensible heat content can decrease, the converse being true for sinking parcels. This is observed in the atmosphere, and it is found that the temperature change is about 0·98°C per 100 m, which is known as the dry adiabatic lapse rate.

Pressure decreases with height in the atmosphere and this can also be used to explain the dry adiabatic lapse rate. If a parcel of air rises, it expands because of the lower environmental pressure, and since the work done by the parcel in so expanding must be at the expense of its internal energy, its temperature falls despite the fact that no heat leaves the parcel. Conversely, the internal energy of a falling parcel is increased and its temperature raised, as a result of the work done on the air in compressing it.

If water-vapour is present in the rising air parcel, a

Figure 3.1 Adiabatic ascent of an isolated air parcel. The rising air parcel cools at the dry adiabatic lapse rate (DALR) until it becomes saturated with water vapour, that is when the dew point and the air temperature become equal. Above the condensation level the air still cools at the DALR, but condensing water vapour releases latent heat and thus the net cooling is less than the DALR, and is known as the saturated adiabatic lapse rate. Above the condensation level the rising air parcel becomes visible because of the water drops it contains. A good approximation to this process can be observed in thermals and cumulus clouds. Cumulus clouds are formed by the condensation of water vapour in thermals of warm air rising from the ground. These clouds have level bases which mark the condensation level in the thermal.

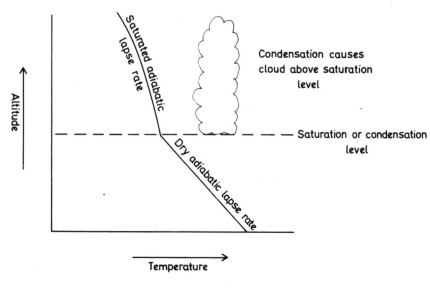

stage will be reached when the air becomes saturated and condensation occurs, thus releasing latent heat. The released latent heat becomes sensible heat and therefore decreases the net lowering of the sensible heat content of the rising parcel. The net result is that the temperature change with increasing altitude is less than the dry adiabatic lapse rate, and is at a rate known as the saturated adiabatic lapse rate. Since the rate of condensation of water-vapour varies with temperature, the saturated adiabatic lapse rate does not have a fixed value, but varies from near that of the dry adiabatic lapse rate for cold air to considerably less for very warm air.

Observation shows that it is justifiable to treat vertically moving individual masses of air as isolated systems which move through the atmosphere without unduly disturbing it or exchanging heat with it. As widespread vertical motion occurs in the lower atmosphere, the average lapse rate of temperature with height is between the dry and saturated adiabatic lapse rates and averages $0.6°C$ per 100 m. Naturally various non-adiabatic processes such as condensation, evaporation, radiation and turbulent mixing also operate to produce temperature changes in the lower atmosphere, but their effects are generally negligible in comparison with those caused by appreciable vertical motion.

Kinetic energy ($V^2/2$) forms a very small part of the total energy content of the atmosphere, amounting to only about 0.5 per cent of the total in the regions of strongest wind speed and considerably less elsewhere. The bulk of the atmospheric energy content is contained in the form of sensible heat plus potential energy, and in normal calculations the energy of the winds is so small that it can be neglected.

The equation for the total energy content of a unit mass of air can be used to study the energy balance of the tropical atmosphere. Outside of southern Asia, the mean north-south circulation of the tropical atmosphere can be considered as taking the form of two simple cells, with rising air near the equator and sinking air over the sub-tropical deserts. The low-level circulations of these cells form the north-east and south-east trade winds. Sinking air in the sub-tropics increases its temperature according to the dry adiabatic lapse rate, thus resulting in clear skies and low relative humidities. Sub-tropical deserts are largely a result of atmospheric subsidence leading to cloudless and rainless conditions.

Large areas of the sub-tropics consist of ocean, and the clear skies result in a plentiful supply of solar radiation reaching the surface where it is mostly used to evaporate sea water. Therefore, over the sub-tropical oceans radiant energy from the sun is turned partly into sensible heat which warms both the atmosphere and ocean, but mostly it is converted into latent heat. There is no evaporation in the dry sub-tropical deserts, so all the incoming radiation is turned into sensible heat resulting in very high day-time temperatures. Radiation losses are also high in deserts, and at night temperatures fall to low values, so the net gain in the sensible heat content of the atmosphere is often small. Water-vapour evaporated over the sub-tropical oceans is mixed through the lower layers of the atmosphere by turbulence and convection and carried towards the equator by the trade winds. Near the equator, the trade winds enter the equatorial trough and here ascent takes place in localized weather systems and in particular in thunderstorms. In the thunderstorms, the latent heat released by the condensing water-vapour is converted into sensible heat which in turn is transformed into potential energy by the rising air mass. In this way, the total potential energy (sensible heat + potential energy) of the rising air in the thunder-cloud is increased by the release of latent heat, and is then exported at high levels in the atmosphere into the sub-tropics and also into middle latitudes.

The atmosphere is not therefore warmed directly by solar radiation, for indeed in the sub-tropics air is actually sinking over the regions of highest surface temperatures, which often decrease towards the equator. There exists instead, a complex mechanism whereby solar energy is turned into latent heat which is transported into the equatorial trough and there converted into sensible heat, so that although the main input of solar energy is in the sub-tropics, the main heating of the atmosphere occurs near the equator. Regions of intense convection with associated clouds in the equatorial atmosphere can be regarded as sites where large amounts of sensible heat are being released and the atmosphere is actually being warmed. It has been calculated that only about 1 500—5 000 active thunderstorms are required to maintain the heat budget of the equatorial trough and thus provide for most of its poleward energy export.

In the atmosphere heat is transported largely by the process of advection, that is by means of the

*Figure 3.2 Energy conversions
in a simple tropical circulation
cell (Hadley Cell).* This type of
circulation cell is observed in
the tropics outside the region of
southern Asia. Short-wave radiation evaporates water vapour
from the oceans and this is
mixed into air flowing towards
the equator at low levels (Trade
Winds), thus increasing its total
energy content. In the equatorial trough the water vapour is
condensed in intense thunderstorms and the latent heat
released coverted into total
potential energy (sensible heat
plus potential energy) which is
exported to the subtropics at
high levels. In the subtropics, air
subsides in anticyclones and
potential energy is converted
into sensible heat. Lastly, the air
continually loses energy in the
form of infra-red radiation.

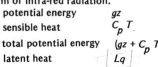

potential energy	gz
sensible heat	$C_p T$
total potential energy	$(gz + C_p T)$
latent heat	$\|Lq\|$

gz + C_pT → Long-wave radiation

$gz → C_p T$

$Lq → C_p T + gz$
Ascent in
thunderstorms

Short-wave radiation → Lq + C_pT   Rainfall

Subtropical   Solar radiation used to warm air and   Equatorial
anticyclone   evaporate ocean water   trough

atmospheric motion, and this is true for both the
sensible and latent heat components. By contrast, in a
solid where there is no motion, the transfer of heat
can only be by the process of conduction. Conduction of heat also occurs in liquids and gases, but it is
normally so slow as compared with the turbulent
mixing process that it is of little importance. The
transfer of heat in a solid can be considered in terms
of an expression which is applicable to many cases
and not just to heat flow:

Input = Storage + Outflow

If the storage is constant or alternatively negligible,
the input of heat would equal the output and the
amount of heat energy (Q) which flows by conduction through the solid during time (t) would be proportional to the cross-sectional area (A) through
which flow takes place, to the temperature difference
(dT) between the two ends of the path, and inversely
proportional to the length of the path (L). That is

$$Q = K \cdot A \cdot dT \cdot \frac{t}{L},$$

where K is a constant of proportionality known as
the thermal conductivity, since it is a physical charac-

teristic of the material through which flow takes
place.

Heat storage is given by the product of the specific
heat and the mass. It is also possible to consider the
thermal capacity of a substance, and this is the product of the density and the specific heat, the units
being cal cm^{-3} deg^{-1}. The soil can be considered as a
series of layers, each of which has a distinctive thermal capacity, temperature gradient and thermal conductivity. The rate at which heat absorbed at the soil
surface flows into the soil will be controlled by all the

Figure 3.3 Flow of heat through a metal bar. If the heat
storage in the bar is constant, flow of heat energy Q during a
given time by conduction is proportional to the cross-sectional area A through which flow takes place, to the temperature difference, $T_H - T_C$, between the two ends of the bar,
and inversely proportional to the length L.

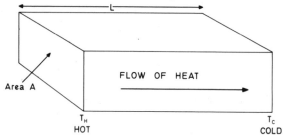

above factors, so some heat will always pass through each layer to the layer below and some will remain in each layer and raise its temperature. The thermal conductivity of soil is low, so heat is conducted through it only very slowly. Warmth applied at the surface only penetrates to the deeper soil layers after a long interval of time, and so temperature changes at depth are small and usually lag behind those at the surface. The maximum temperature at a depth of 20 cm is normally reached several hours after that at the surface. Since heat only penetrates slowly into the soil, the temperature changes at the soil surface are large and follow closely the diurnal and seasonal variations in solar radiation. In the oceans heat is mixed downwards by turbulent motions in the water, a much more effective process than simple conduction and so temperature changes at the sea surface are always small. The different reactions of land and water surfaces to solar radiation are further accentuated by the fact that the specific heat of soil is much smaller than that of sea water. So soil surface temperatures follow changes in solar radiation closely but temperature changes at depth are small, while sea surface temperatures tend to remain constant and only follow changes in radiation very slowly.

Bernoulli's Theorem

This theorem states that in a fluid in steady motion and without friction the sum per unit mass of the kinetic energy ($V^2/2$), the potential energy possessed by virtue of being in a pressure field (p/ρ), and the gravitational potential energy (gz) is constant:

$$\frac{V^2}{2} + \frac{p}{\rho} + gz = \text{constant},$$

where p is the pressure,
         ρ the density,
and the other terms are as previously defined.

Bernoulli's equation may be simplified in a number of ways. Thus if a small element of air passes at a constant altitude above sea-level from high to low pressure, the kinetic energy will increase by an amount proportional to the change in pressure, i.e.

$$\frac{V^2}{2} + \frac{p}{\rho} = \text{constant}.$$

Similarly, if the particle moves from low to high pressure it will eventually come to rest when the second term in the above equation and the constant become equal.

When the density of the fluid is uniform and there is no motion the pressure decreases uniformly with height and

$$\frac{p}{\rho} + gz = \text{constant}.$$

A third example is provided by a gliding bird in still air, where if friction is neglected, Bernoulli's equation becomes:

$$gz + \frac{V^2}{2} = \text{constant}.$$

When the bird dives its speed increases by an amount proportional to the loss in gravitational potential energy, while if the bird rises its speed falls.

The flow of water in a river channel can be considered in terms of the last example. As the water flows to lower elevations it should accelerate in velocity, but most of the energy released by the decrease in gravitational potential energy is used in overcoming the resistance to water movement provided by the channel. Energy used in this way appears as sensible heat, but the rise in temperature of the water is very small.

4 Radiation in the Tropical Environment

ANY object not at a temperature of absolute zero (−273°C) transmits energy to its surroundings by radiation, that is, by energy in the form of electromagnetic waves travelling with the speed of light and requiring no intervening medium. This radiation is characterized by its wave-length, of which there is a wide range or spectrum extending from the very short X-rays through the ultraviolet and visible to infra-red, micro-waves and radio waves.

A valuable theoretical concept in radiation studies is that of the blackbody, which is one that absorbs all the radiation falling on it and which emits, at any temperature, the maximum amount of radiant energy. The term arises from the relation between darkness of colour and the proportion of visible light absorbed, since a body that appears white scatters most of the visible light falling on it. For a perfect all-wave blackbody, the intensity of radiation emitted and the wave-length distribution depend only on the absolute temperature, and in this case a number of simple laws apply. The Stefan-Boltzmann law states that the amount of energy (F) emitted in unit time from a unit area of a blackbody is proportional to the fourth power of its absolute temperature (T), i.e.

$$F = \sigma T^4,$$

where σ is Stefan's constant.

Thus the higher the temperature of an object the more radiation it will emit.

As shown in the accompanying diagram, a blackbody does not radiate the same amount of energy at all wave-lengths for any given temperature. At a given temperature, the energy radiated reaches a maximum at some particular wave-length and then decreases for longer or shorter wave-lengths. The Wien displacement law states that this wave-length of maximum energy (λ_{max}) is inversely proportional to the absolute temperature, i.e.

$$\lambda_{max} = \frac{\alpha}{T}$$

where α is a constant.

Thus as the temperature of an object increases, the wave-length of maximum energy decreases, passing from the infra-red for objects at room temperature to the visible wave-lengths for extremely hot objects.

If the sun is assumed to be a blackbody, then an estimate of its effective radiating temperature may be obtained from the Stefan-Boltzmann law, which suggests an effective surface temperature of 5 750°K. For the sun, the wave-length of maximum emission is near 0·5 μm ($10^{-6}m$), which is in the visible portion of the electro-magnetic spectrum, and almost 99 per cent of the sun's radiation is contained in the so-called short wave-lengths from 0·15 to 4·0 μm. Observations show that 9 per cent of this short-wave radiation is in the ultraviolet (less than 0·4 μm), 45 per cent in the visible (0·4 to 0·7 μm) and 46 per cent in the infra-red (greater than 0·74 μm).

The surface of the earth, when heated by the absorption of solar radiation, becomes a source of long-wave radiation. The average temperature of the earth's surface is about 285°K (12°C), and therefore most of the radiation is emitted in the infra-red spectral range from 4 to 50 μm, with a peak near 10 μm, as indicated by the Wien displacement law. This radiation may be referred to as long-wave, infra-red, terrestrial or thermal radiation.

It is essential to distinguish clearly between reflected and re-radiated radiation. If radiation is directly reflected there is no change in wave-length, and so short-wave solar radiation is reflected as short-wave radiation. Some of the short-wave radiation falling on

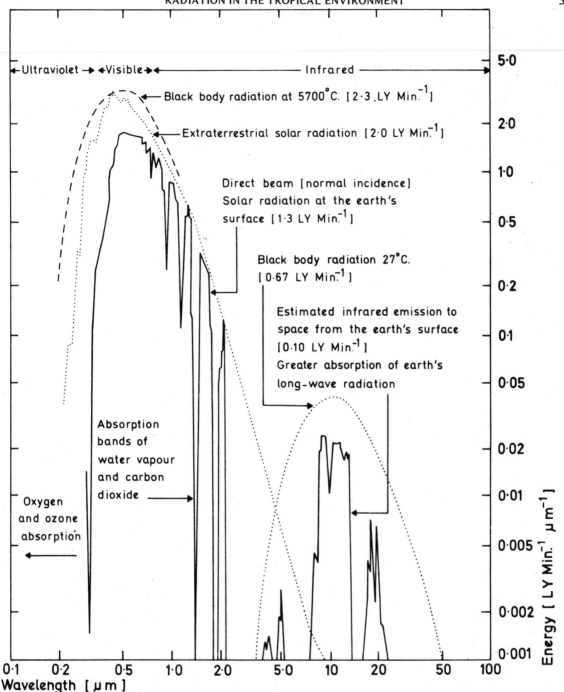

Figure 4.1 Radiation from the sun and the earth. The horizontal scale shows the wavelength in μm $(10^{-6}m)$, while the vertical scale indicates the energy at that wavelength in terms of cal/cm²/min/μm. Ultraviolet, visible and infra-red radiations are shown at the top of the diagram. The radiation emission from a black body at 27°C is also shown, and it is clearly seen that the energy radiated reaches a peak at a particular wavelength and then declines on either side of the peak. The black body radiation at 5,700°C is assumed to be that from the sun at this temperature, and is reduced so that it is the radiation which would be received at the top of the earth's atmosphere. The absorption of infra-red radiation by water vapour and carbon dioxide is clearly illustrated. (After Sellers, 1965). (Reproduced by permission of University of Chicago Press).

TABLE 4.1

VALUES OF ALBEDO FOR VARIOUS SURFACES

Surface	Albedo (percentage of incoming short-wave radiation which is reflected)
Deciduous forest	17
Desert scrubland	20–29
Swamp	10–14
Ploughed field, moist	14
Sand, bright, fine	37
Dense, dry clean snow	86–95
Woody farm covered by snow	33–40

TABLE 4.2

MEAN PLANETARY TEMPERATURES

Planet	Mean Distance from Sun (Earth = 1)	Mean Planetary Albedo (percentage)	Mean Temperature (°K)
Mercury	0·387	6	616
Venus	0·723	76	235
Earth	1·000	30	254
Mars	1·524	16	209
Jupiter	5·20	73	105
Saturn	9·55	76	78
Uranus	19·2	93	55
Neptune	30·1	94	43
Pluto	39·4	14	42

Mean distances are given as a multiple of earth's mean distance from the sun.

The mean temperature is that temperature which is required, with the given albedo, for the planet to emit to space as infra-red radiation exactly the same amount of energy as it receives as solar radiation from the sun.

the surface of the earth is reflected, and the ratio of the incoming to the reflected surface short-wave radiation is known as the albedo. Some typical values of albedo are shown in the accompanying table, and clearly the albedo of a blackbody is zero. If the radiation is absorbed by the surface and then re-radiated, the wave-length of the re-radiated radiation will vary according to the Stefan-Boltzmann and Wien laws, that is, it will be controlled by the absolute temperature of the surface. Often, the temperature of the surface will adjust itself, so that as much energy is being radiated as is being absorbed, and thus the body will neither gain nor lose energy in the long term. The mean surface temperature of the earth is such that the earth radiates as much energy to space as it receives from the sun, so the long-term energy content of the planet is almost constant. In the last example it should be remembered that the intensity of solar radiation is greatly reduced with distance from the sun, and therefore the further planets are from the sun, the lower the surface temperature required to achieve radiation balance. Therefore planets near the sun, such as Mercury, are intensely hot, while distant planets, such as Neptune, are extremely cold.

From the discussion so far it will have become clear that the radiation experienced in the natural environment has two main components, short-wave and long-wave radiation, and that radiation studies are very much concerned with the individual study of these two components and also with their interactions.

Short-Wave Radiation

Even under cloudless skies, not all the solar radiation falling on the outer limits of the earth's atmosphere penetrates through to the ground surface. In an atmosphere which contains neither water-vapour nor dust, the depletion of the solar radiation is mostly by molecular scattering with only a very small amount of energy being directly absorbed by the air. The radiation falling on the ground can therefore be divided into two components—the first being the direct solar radiation from the sun and the second the scattered radiation from the sky. The amount of scattering is inversely proportional to the wave-length of the light, so it is greatest in the ultraviolet and least in the infra-red, and this is why the sky appears blue. Dust and smoke greatly increase the amount of scattering and absorption and thereby reduce the intensity of the direct radiation at the earth's surface, which even under cloudless skies will vary slightly with the air mass, and in particular will be least in smoky cities and in dusty arid areas.

If clouds are present in the atmosphere, then the distribution of radiation at the surface will be further complicated because not only will the amount of direct radiation be reduced, but the clouds may also increase the amount of diffuse short-wave radiation from the sky. Clouds have two basic effects on the incoming radiation, firstly they reflect solar radiation directly back to space and this represents a complete loss of energy to the earth, and secondly some radiation diffuses through the cloud to provide general uniform illumination on the surface below. The exact amount of diffuse radiation reaching the surface will depend on the amount, height and thickness of the cloud.

Global radiation is the sum of both the direct and the diffuse radiation falling on a horizontal surface, and its intensity depends on both the angle of the sun above the horizon and the amount of cloud in the atmosphere. Direct solar radiation from the sun is direction-dependent and increases in intensity as the sun gains in altitude above the horizon, but in contrast, diffuse radiation can be assumed to be independent of direction and of nearly equal strength in all directions.

To calculate the direct component of the global radiation it is necessary to know the position of the sun in the sky, and this is controlled by the planetary and rotational motions of the earth. The earth moves round the sun in a near-circular orbit in very slightly

less than 365 days, while at the same time it spins once in 24 hours on an axis set at an angle of 66° 33' to the plane of its orbit. The axis of rotation remains fixed in space, pointing in the north to within a degree of Polaris and in the south to an unmarked position in the heavens about 28° below the foot of the Southern Cross. Thus each pole is inclined towards the sun and away from the sun alternatively once per year as the annual planetary motion proceeds.

The celestial sphere is an imaginary sphere, concentric with the earth, on the inner surface of which the heavenly bodies (sun, stars etc.) appear to lie, the observer being situated at the centre of the sphere. For an observer at the centre of the celestial sphere, the celestial poles are points on the celestial sphere in the direction parallel to the earth's axis of rotation, and as already noted they are fixed relative to the distant stars. It therefore follows that the celestial equator is that great circle whose plane is perpendicular to the line joining the celestial poles. The angular distance of a heavenly body north or south of the celestial equator is known as its declination, and the declination of the sun varies throughout the year because of the rotation of the earth around the sun and the inclination of the axis to the orbit. It varies from about 23° 27'N (90°–66° 33') at the June solstice,

Figure 4.2 Rotation of the earth about the sun and solar declination. Since the earth's axis of rotation is tilted at an angle of 66° 33' to the plane of its orbit around the sun, solar declination varies from 0 degrees at the March and September equinoxes to 23° 27'N and S at the June and December solstices respectively.

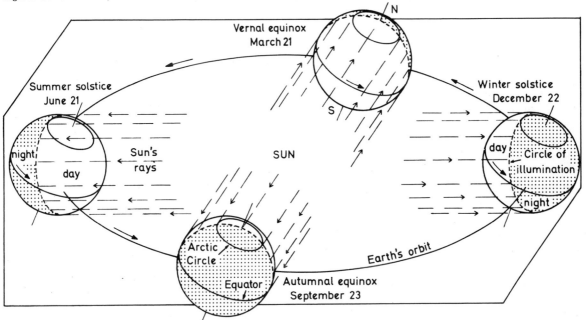

through 0° at the March and September equinoxes, to about 23° 27'S at the December solstice.

The solar declination is independent of the position of the observer on the earth's surface since it depends only on the time of year, and it may therefore be obtained by reference to a Star or Nautical Almanac, or it may be calculated directly on a computer. The angle of the sun above the horizon at local noon on any day is given by:

$$\text{angle of sun above horizon} = 90° - \text{latitude of observer} + \text{solar declination}$$

A few simple calculations show that the solar altitude will vary progressively throughout the year, being greatest in summer and least in winter in the regions poleward of the tropics (23° 27' N and S), while poleward of 66° 33' N or S the sun never rises above the horizon on mid-winter day. Between the two tropics the sun is directly overhead at local noon on two occasions each year, and at the tropics themselves it will be directly overhead at local noon on mid-summer's day.

As the angle of the sun above the horizon at local noon varies so also does the length of daylight at any given latitude. The variation in the length of daylight is greatest at the poles and least at the equator, since the mid-summer pole receives 24 hours of daylight while the mid-winter pole has 24 hours of darkness. At the equinoxes the length of daylight is everywhere about 12 hours, after which it changes rapidly to reach the maximum difference at the solstices. Near the equator the length of daylight is always about 12 hours and its variations are not significant.

It will be clear by now that the daily totals of global radiation will be influenced not only by the solar altitude above the horizon but also by the length of daylight. The distribution of radiation over the earth's surface is considered later, but it should already be clear that it will show large seasonal variations.

Even under completely cloudless conditions some solar radiation is lost during the passage through the atmosphere, and this loss is partly proportional to the angle of the sun above the horizon, since the lower the sun the longer the path of the solar beam through the atmosphere. The depletion of the solar beam varies in different parts of the world, but for the clear air of the Atlantic Ocean it can be estimated from:

$$F = 0.6 + 0.2 \sin \theta,$$

where F is the fraction of the solar radiation transmitted through the atmosphere,

and θ is the angle of the sun above the horizon. If the sun is overhead then $\sin \theta$ is equal to one ($\sin 90° = 1$) and 80 per cent of the incident solar beam reaches the surface, but when the sun is only just above the horizon then $\sin \theta$ is equal to zero ($\sin 0° = 0$) and only 60 per cent of the incident radiation reaches the surface. The intensity of the incident beam (S) on a horizontal surface is given by:

$$S = (\text{solar constant}) \times \sin \theta$$
$$(\text{where } \theta \text{ is the solar angle})$$

and the solar constant is defined as the solar radiation falling on a surface normal to the sun's beam outside of the earth's atmosphere, at the earth's mean distance from the sun. The generally accepted value of the solar constant is $139.6 \text{ m } W \text{ cm}^{-2}$ or $2 \text{ cal cm}^{-2} \text{ min}^{-1}$.

Cloud causes further depletion of the solar radiation reaching the surface, but the amount of attenuation depends very much on the nature of the cloud and its amount. Thus high cloud causes less depletion than dense low cloud, as observation shows. The influence of cloud on incoming radiation may be estimated from the following relation:

$$\text{Depletion} \propto F(1 - 0.4C_h)(1 - 0.7C_m)(1 - 0.7C_\ell),$$

where F is the fraction of the solar beam which would reach the surface if there were no cloud present,

and C_h, C_m and C_ℓ are the percentages of the sky covered by high, medium and low cloud respectively. It is seen that the amounts of medium and low cloud are most significant in depleting the solar radiation. If the sky were completely covered by low cloud the solar radiation reaching the surface would be reduced to 30 per cent of the original value, while a half cover of low cloud would reduce it to 65 per cent of the clear-sky value.

So far it has been assumed that radiation falls on a horizontal surface, but natural surfaces are at many angles which are not usually near the horizontal. If the surface is not horizontal then the intensity of radiation may be increased or decreased depending on whether the surface is tipped towards or away from the direction of the solar beam. A surface with a slope in a north-south direction may be considered as equivalent to the surface undergoing a change in latitude, since a slope towards the equator implies a decrease in latitude while a slope towards the pole implies an increase. Similarly, a tilt towards the east or west is equivalent to a change in longitude. Normally slopes tilted towards the equator receive more radiation than horizontal surfaces and therefore have an advantage of relatively higher temperatures and longer growing seasons. The difference between radiation amounts received by various slopes at a given latitude may not be as great as first expected because of the diffuse component in the radiation. The assumption in all calculations of radiation falling on slopes is that the incident radiation consists mostly of direct solar radiation, which is direction dependent. This is true if the atmosphere is cloudless, but as the

amount of cloud increases so also does the proportion of diffuse radiation in the global radiation, and thus in cloudy weather radiation may be almost equal on all slopes. In the tropics, where the sun is always high in the sky at noon, nearly all slopes receive an equal amount of radiation over the year, so cloud amount is not particularly relevant. Skies in the polar regions are often covered by thin cloud, resulting in a high percentage of diffuse radiation, and this coupled with the fact that the sun tends to circulate around the horizon in summer, makes small differences in slope unimportant as regards the radiation climate. Angle of slope appears to be most important in middle latitudes $(40°-60°)$, where the resulting difference in radiation received can have a marked influence on vegetation.

It is of some interest to consider the loss of radia-

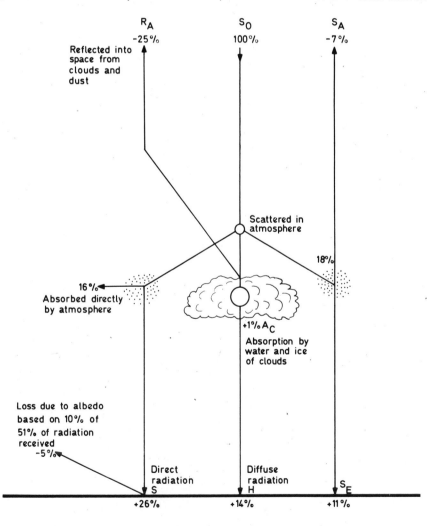

Figure 4.3 Interactions of short-wave radiation with an average atmosphere. It is assumed that 100 units of short-wave radiation are incident on the top of the atmosphere, and it is then possible to express all interactions in terms of percentages of the original incoming radiation. Downward fluxes of radiation are positive while upward fluxes are negative. Diffuse radiation reaching the ground is made up of $H + S_E$ (25 per cent) which is almost equal to the direct radiation S (26 per cent). $S + H + S_E$ together form the global radiation falling on a horizontal surface. (After Flohn, 1969).

tion in an atmosphere possessing properties which are average for the earth as a whole, that is it contains an average amount of cloud, dust, etc. One such estimate is shown in the accompanying diagram, which is an average for the whole year. In this diagram it is assumed that the incident radiation is equivalent to 100 units and then all other interactions can be expressed as a percentage of the original input. Dust and haze in the atmosphere scatter on average 18 per cent of the extra-terrestrial incoming radiation, and of this amount 11 per cent reaches the earth, while 7 per cent is scattered back into space. Clouds in the atmosphere, which receive about 40 per cent of the incoming radiation immediately reflect 24 per cent back to space, while of the remainder a small proportion (2 per cent) is absorbed by water droplets and ice particles, and 14 per cent forms diffuse radiation which reaches the earth's surface. The amount of direct radiation reaching the surface of the earth is thus only about 26 per cent of the original incoming solar radiation, but to this must be added the diffuse radiation from the sky (11 per cent) and from clouds (14 per cent), making a total for the global radiation of 51 per cent of the original input. The exact amount of the global radiation which is reflected depends on the albedo of the surface which averages 10 per cent but can reach 80 per cent over snow, and this reduces the amount of extra-terrestrial radiation absorbed by the soil surface to 47 per cent. This value of 47 per cent absorbed by the soil surface should be compared with the 16 per cent absorbed directly by the atmosphere, indicating that the atmosphere is largely transparent to short-wave radiation, the bulk of which is absorbed by the ground. This result is of great importance, since it indicates that the atmosphere is heated from below, and the results of this are of such fundamental importance that they will be discussed again later.

Long-Wave Radiation

There are two sources of long-wave radiation near the ground surface; the first originates from the surface itself and the second comes from the atmosphere. The earth's surface is usually assumed to emit and absorb in the infra-red region as a grey body, that is, as a body for which the Stefan-Boltzmann law takes the form:

$$F = \epsilon \sigma T^4,$$

where the constant of proportionality ϵ is defined as the infra-red emissivity or, equivalently the infra-red absorptivity (one minus the infra-red albedo). Clearly, the emissivity of a blackbody is one, and it appears that practically all natural surfaces, including snow, have emissivities of between 90 and 95 per cent, corresponding to infra-red albedos of 10 to 5 per cent. The wave-length distribution of the radiation from natural surfaces is controlled by the absolute temperature according to the Wien displacement law, which was discussed in the introduction to this chapter, and for the range of temperatures found on the earth's surface, lies between 4 and 50 μm with a peak near 10 μm.

The atmosphere is nearly transparent to short-wave radiation from the sun, of which large amounts reach the earth's surface, but it readily absorbs infra-red radiation emitted by the earth's surface, the principal absorbers being water-vapour (5·3 to 7·7 μm and beyond 20 μm), ozone (9·4 to 9·8 μm), carbon dioxide (13·1 to 16·9 μm) and clouds (all wave-lengths). Only about 18 per cent of the infra-red radiation from the ground surface escapes directly to space, mainly in the atmospheric window between 8·5 and 11·0 μm, the rest being absorbed by the atmosphere, which in turn re-radiates the absorbed infra-red radiation, partly to space and partly back to the surface. The main receiver of heat from the sun is thus the ground surface, while most of the heat lost to space is radiated from the lower troposphere. There must therefore be a continual passage of energy from the surface to the atmosphere, and the nature of this transformation was discussed in the previous chapter.

The general level of temperature in the lower troposphere has to be such that the earth as a whole radiates, in the long term, as much energy to space as it receives from the sun. Now if heat is to flow from the ground surface to the lower atmosphere there must be a suitable temperature gradient, and this suggests that the general temperature of the ground surface will be above that of the lower atmosphere. The earth's surface is thus at a higher average temperature than that which would be observed if the atmosphere were perfectly transparent or absent. Indeed, it has been estimated that if the atmosphere were absent, the earth's surface would be 30° to 40°C cooler than it is at present. The term 'greenhouse effect' is often applied to this particular phenomenon, because it is

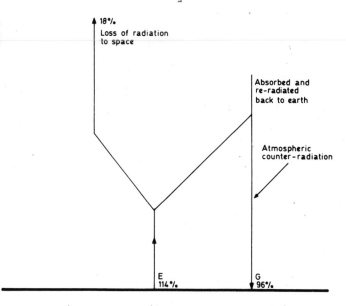

18%
Loss of radiation
to space

Absorbed and
re-radiated
back to earth

Atmospheric
counter-radiation

E
114%

G
96%

Figure 4.4 Interactions of long-wave radiation with an average atmosphere. The units are percentages of the initial incoming short-wave radiation S in Figure 4.3, and so Figures 4.3 and 4.4 are directly comparable. E represents the infra-red radiation from the ground surface most of which is absorbed by water vapour, carbon dioxide and cloud in the atmosphere, and radiated back to the ground as the atmospheric counter-radiation G. (After Flohn, 1969).

analogous to the effect which is supposed to operate in a greenhouse where the glass is transparent to short-wave radiation but nearly opaque to infra-red radiation.

An annual budget of the long-wave terrestrial radiation of the earth is shown in the accompanying diagram, the percentages being in terms of the original incoming extra-terrestrial radiation. The emission of radiation by the earth's surface, calculated from the average temperature, averages about 114 per cent of the incoming extra-terrestrial radiation. Fortunately, the greater part of this emission is absorbed by the water-vapour and carbon dioxide in the lower layers of the atmosphere and re-radiated. The infra-red energy which is radiated downwards is known as the atmospheric counter-radiation, and has an average value of about 96 per cent of the solar constant. Because of this continual interchange of radiation between the earth's surface and the lower atmosphere, the effective loss of energy from the ground surface is only about 18 per cent of the solar constant, a value which varies very little with the season or latitude. Observations show that the lowest 100 m of the atmosphere contribute strongly to the atmospheric counter-radiation reaching the surface and that each 100 m above has rapidly diminishing influence, so that conditions above 1 000 m are usually of little significance with a clear sky.

When considering the effect of cloud on atmospheric counter-radiation there are two main com-ponents which are cloud amount and the temperature of the cloud base, and since high level clouds have lower base temperatures than low clouds it therefore follows that cloud height is important.

The Net Radiation

Net radiation is the difference between the total incoming and total outgoing radiation, and clearly it shows whether net heating or cooling is taking place. It will normally be negative at night indicating cooling but it may be negative or positive during the day, depending on the balance of the incoming and the outgoing radiation.

The accompanying diagrams illustrate the components of the net radiation balance, assuming that there is no evaporation nor condensation. The simplest case occurs at night, because there is no incoming short-wave radiation, but instead there is a continuous long-wave radiation loss. At night the soil surface emits long-wave radiation to the atmosphere, where water-vapour and carbon dioxide absorb large amounts of it, which in turn is partly re-radiated downwards to be re-absorbed by the soil surface. The difference between the two streams represents the net radiation loss, which in this example represents a cooling and consequently there is a flow of sensible heat from the atmosphere and the lower layers of the soil towards the soil surface.

During the day the situation is more complex because of the incoming short-wave radiation, which leads to interactions additional to those described for the night. Short-wave radiation reaching the surface can be either absorbed or reflected. The reflected radiation represents a complete energy loss, because it does not heat the surface and it must be added to the negative terms in the net radiation equation. Of the short-wave radiation absorbed, some is re-radiatèd as long-wave radiation and some is transferred as sensible heat into the soil and the atmosphere.

Energy Balance

To fully describe the total energy balance of an object it is necessary to take into account not only radiation but also gains or losses of sensible and latent heats. This is well illustrated by the study of the energy balance of a small object such as a plant leaf. The energy flow between a plant leaf and its environment may be described by the following equation:

Radiation absorbed = long-wave radiation loss $(\in \sigma T_{\ell}^4)$ ± energy gained or lost by convection + energy lost in the form of latent heat (LE_t) ± energy for basic metabolism in the plant;

where $\in$ is the infra-red emissivity,

σ is Stefan's constant,

T_{ℓ} is the leaf temperature,

L is the latent heat of water,

and E_t is the transpiration rate from the leaf surface.

Usually the energy used for metabolism within the plant is small and can be neglected in energy balance calculations.

Absorbed radiation forms the main energy input and this depends on both the incident radiation and the albedo of the leaf surface. The incident radiation consists of short-wave radiation from the sun and the sky together with radiation reflected from the ground and from neighbouring leaves and plants. Thus short-wave radiation will fall on both the upper and lower surfaces of the leaf. To the short-wave radiation should also be added a long-wave radiation gain, consisting partly of infra-red radiation from the ground and nearby plants and partly of atmospheric counter-radiation. In the far infra-red, plants act very nearly as blackbodies, so about 95 per cent of the long-wave radiation will be absorbed by the leaf.

Long-wave radiation loss depends on the absolute temperature of the leaf. If there were no other sources of energy loss, then the temperature of the leaf would adjust itself so that the same amount of

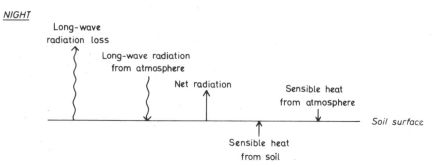

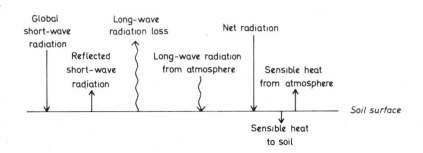

Figure 4.5 Net radiation over a dry soil surface. Downward fluxes of radiation are positive, while upward fluxes are negative. The net radiation is obtained by the addition of all the radiation fluxes having regard to sign, so a negative net radiation indicates a cooling of the soil surface while a positive indicates a warming. At night there is net cooling, so sensible heat flows towards the soil surface, while during the day there is net warming with sensible heat flowing away from the surface.

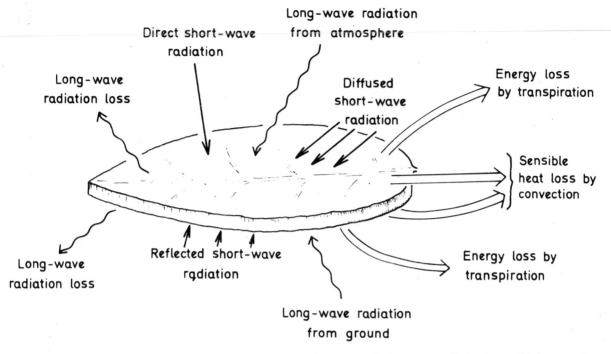

Figure 4.6 The energy balance of a plant leaf. The leaf gains energy by absorbing both short and long-wave radiation. It loses energy by long-wave radiation, by sensible heat transfer to the atmosphere and by latent heat loss.

radiation is lost as is gained. This would occur through the action of the Stefan-Boltzmann radiation law. Since energy is also lost by convection and evapotranspiration, the surface temperature of the leaf is reduced below that which it would have been, if it were in pure radiative balance.

Sensible heat loss by convection (C) depends on the temperature difference between the leaf (T_{ℓ}) and the air (T_a), the wind speed (V) and also the shape and size of the leaf, which may be represented by the width (D):

$$C \propto \frac{V^{\frac{1}{2}}}{D} (T_{\ell} - T_a)$$

Thus the greater the temperature gradient between the leaf and the air, the more rapid will be the sensible heat loss. A hot leaf in a cold atmosphere will cool more quickly by convective heat loss than a hot leaf in a hot atmosphere. Wind speed is important because the stronger the wind the more quickly is heat carried away from the leaf. Size enters the heat loss relationship because the temperature of a flat leaf which is normal to the incident radiation will be strongly influenced by it, while a cylinder, such as a branch, twig or pine needle, will be much less strong-

ly influenced. Usually, rapid sensible heat losses are associated with small objects and so their temperatures are very close to that of the air.

Latent heat loss depends partly on the wind speed and partly on the relative humidity of the air. Normally the relative humidity within the stomata of plant leaves will be nearly 100 per cent, and thus there will always be some water loss to the atmosphere whenever the stomata are open. Water loss to the atmosphere may be limited because the stomata are closed or because the resistance to the passage of water-vapour from the leaves to the atmosphere is very high. In many plants, such as grasses, this resistance to the diffusion of water-vapour from the leaf to the air is small and may be neglected. In others, such as conifers, it is large and therefore causes the actual transpiration from the leaves to be markedly below that of such plants as grasses under similar conditions.

If infra-red radiation were the only form of heat loss, the temperature of plant leaves would be completely controlled by the radiation climate. Sensible heat loss by convection and latent heat loss tend to reduce the influence of the incident radiation, that is to say they uncouple the temperature of the leaf

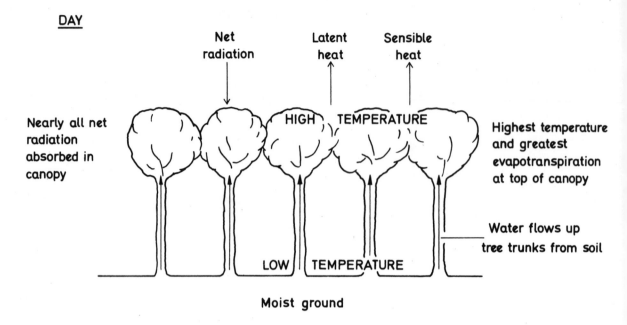

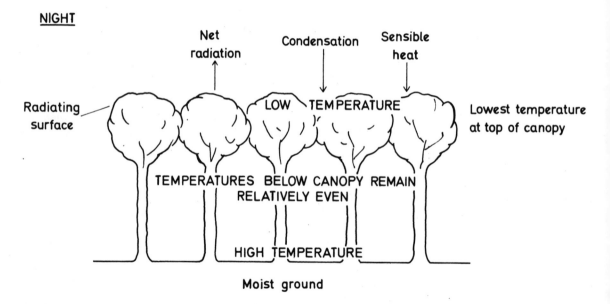

Figure 4.7 Energy exchanges within a forest.

Upper. During the day the top of the tree canopy gains heat energy from the downward flux of net radiation. The canopy loses heat energy by warming the atmosphere (sensible heat loss) and by evapotranspiration (latent heat loss). Since most of the net radiation is absorbed by the top of the canopy, most of the evapotranspiration and sensible heat loss occur at this particular level. Water for evapotranspiration flows up the tree trunks from the moist ground. While temperatures in the tops of the trees is high, the air temperature near the ground remains relatively low.

Lower. At night energy is lost from the top of the canopy by the upward flux of net radiation which cools it to temperatures below those found near the ground. Energy is supplied to the top of the canopy by a flow of sensible heat from the air and perhaps also by condensation in the form of dew. While temperatures at the top of the canopy follow the diurnal variation of net radiation, those near the soil surface remain relatively constant.

from the radiation environment. The smaller the leaf the more effectively it is uncoupled from the incident radiation and the more nearly will its temperature approach that of the air.

There is a tendency for the difference between leaf and air temperatures to be positive in cool weather, and negative in hot. Observations indicate that the cross-over temperature for well-watered, thin-leaved plants exposed to noonday sunshine is about 30°C. Some research workers consider that the optimum temperature for the growth of many plants is near 31°C, and so leaf anatomy and morphology would evolve to bring leaf temperatures towards their optimum at the time of day when most sunlight is available for photosynthesis. It has been found that the upper limit for air temperatures just above well-watered thin-leaved vegetation is about 33°C. This is because air temperatures can only rise by the flow of heat from the leaves, and this will only occur whilst the leaves are warmer than the atmosphere. So in a well-watered completely vegetated region, the maximum air temperatures must be close to the upper limit for leaf temperatures.

Leaf temperatures within a given plant will vary widely because of their differing inclinations to the incoming radiation. Also some leaves within the canopy will be partly shaded by others, and as a result the active surface where most of the radiative exchanges take place will be at the top of the canopy. Thus the upper leaves of the plant will be warmest during the day because this is where the main energy gain occurs, and coolest at night because of the relatively large energy loss. Leaves further within the canopy will experience less marked temperature changes because they are insulated from changes in net radiation. In the case of small plants such as grasses, the active surface is almost coincident with the ground, but in forests the active surface is far above ground level and a separate micro-climate is created within the forest below the top of the canopy. Thus within a dense forest, day-time temperatures are normally lower and night-time temperatures higher than at the same level over nearby short grass. Similarly, it follows that most of the evapotranspiration within a forest takes place from the top of the canopy.

Bowen Ratio

The study of the energy balance of natural surfaces falls into two stages. Firstly, there is the study of the radiation balance, which leads to an estimation of the available net radiation. Secondly, the net radiation has to be divided amongst the sensible heat flows to the soil and the atmosphere, and the latent heat flow to the atmosphere, to produce the full energy balance. The ratio of the sensible heat flow to the atmosphere to the latent heat flow is known as the Bowen ratio, and can be written as:

Bowen ratio (β)

$$= \frac{\text{Sensible heat loss to atmosphere } (K)}{\text{Latent heat loss to atmosphere } (LE_t)}$$

If the surface is saturated and evaporating freely, and if it is in simple radiative balance, that is to say heat is not being advected in by the atmosphere, then the value of the Bowen ratio is largely dependant on the surface temperature. This is clearly seen in Figure 4.8, where the ratio of sensible heat loss to latent heat loss is plotted against temperature. When the surface is dry, evaporation is restricted and part of the net radiation which would have formed latent heat loss now becomes sensible heat, resulting in an increase in the value of the Bowen ratio above that appropriate for a freely-evaporating surface at the same temperature. Similarly, if the evaporating surface is small in area, heat may be imported by the winds from outside and this may be used in evaporating water. Therefore the simple relationship shown in Figure 4.8 only applies to wet surfaces of large extent.

Figure 4.8 suggests that over a well-watered surface the Bowen ratio decreases as the temperature increases, or alternatively the proportion of available energy going into latent heat increases. At temperatures above 32°C the sensible heat flow becomes negative implying a flow of heat from the air to the evaporating surface. This would seem to suggest that the highest temperature which can be reached over a freely-evaporating surface with the net radiation values experienced on the earth's surface is about 32°C. In an earlier part of this chapter it was mentioned that the highest temperature observed over well-watered, thin-leaved vegetation is about 33°C, and this provides another explanation of this observation. Thus the upper limit of temperature under normal conditions in tropical rain forest is about 33°C, and it will only exceed this value if the soil becomes dry under drought conditions. Temperatures in moist

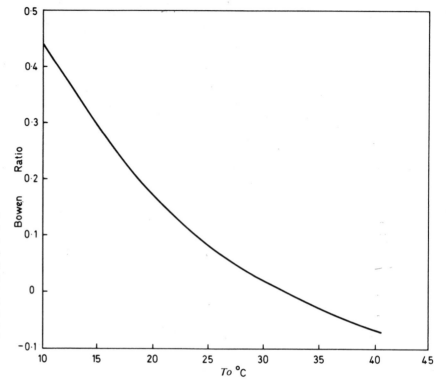

Figure 4.8 Values of the Bowen Ratio at a freely evaporating suface for various surface temperatures (T_o). The Bowen ratio is the ratio of the sensible heat loss to latent heat loss. This diagram only applies to a freely evaporating surface, for if the surface becomes dry, the value of the Bowen ratio will increase to above that appropriate for the temperature T_o.

rain forest will also be relatively insensitive to small changes in net radiation, and are therefore relatively uniform in both space and time.

Under arid conditions the upper limit to temperature is set by the radiation balance of the surface, there being little or no latent heat loss. Thus in dry desert areas, air temperatures may reach 50° or 55°C, since all the net radiation is available for sensible heat transfer to the atmosphere. In tropical climates with alternating wet and dry periods, lower temperatures are often observed in the wet periods partly because the rise in temperature is limited by evaporation.

Radiation Within the Tropics

Global radiation, consisting of direct solar and diffuse sky radiation, is the principal component of the radiation balance and the most important. It is seen from Figure 4.9 that it varies between 120 and 220 kilo-calories/cm²/year over the Indian Ocean and its adjacent land areas, the highest total being observed in the sub-tropics over the deserts of West Asia, Africa and Australia. The maximum yearly totals of 200–220 kilo-calories/cm² are observed in

north-east Africa and Arabia. In contrast, the equatorial zone in general receives annual totals of no more than 150 kilo-calories/cm² and in South-East Asia and equatorial Africa only 120–140 kilo-calories/cm². The surface values of global radiation are affected mainly by cloud amount and to a lesser extent by atmospheric dust, haze and water-vapour. Spatial variations of global radiation thus mainly reflect variations of cloudiness, so it is large over deserts where cloudiness is least and small over regions of greater cloudiness near the equator.

There are marked seasonal variations in global radiation. In January the maximum occurs in the southern hemisphere along 30°S, but the pattern changes gradually as the sun moves north, so that by May the pattern is reversed with high values along 25°N. In June the pattern changes abruptly with the establishment of the south-west monsoon over southern Asia, minimums appearing over western India and western Burma.

Figure. 4.9 illustrates the annual distribution of net radiation. Values over the sea are always higher than those over the land, in contrast to the distribution of global radiation. This is because of variations in

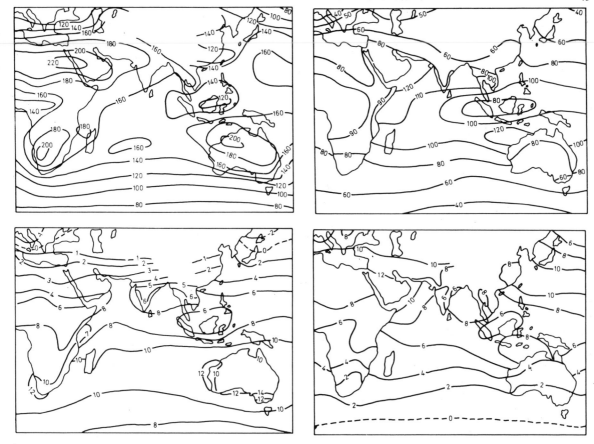

Figure 4.9 Distribution of global and net radiation over the Indian Ocean and neighbouring land areas.

Top left. Annual distribution of global radiation (k. cal. cm^{-2} year^{-1})

Top right. Annual distribution of net radiation (k. cal. cm^{-2} year^{-1})

Bottom left. Amount of net radiation in January (k. cal. cm^{-2} month^{-1})

Bottom right. Amount of net radiation in July (k. cal. cm^{-2} month^{-1})

(After A. Mani, O. Chacko, V. Krishnamurthy, and V. Desikan, 1967: Distribution of global and net radiation over the Indian Ocean and its environment. Archiv Meteorologie Geophysik und Bioklimatologie, Ser B. 15, 82).

albedo and net long-wave radiation loss between land and sea surfaces. The albedo of the sea surface is low, only about 10 per cent, compared to that of the land, which typically has values in the range of 15–25 per cent. Albedo values in dry desert areas are even higher, reaching 30 per cent or more. Also the greatest amounts of net long-wave radiation loss are observed in tropical deserts, where values reach 80 kilocalories/cm^2/year. This is mainly the result of the high temperature of the desert surface and the dryness of the air above. Nearer the equator, the net long-wave radiation loss is only about 30 kilocalories/cm^2/year and varies only slightly between land and sea.

Annual values of net radiation are surprisingly constant over the tropical land masses surrounding the Indian Ocean, values mostly being of the order of 80 kilo-calories/cm^2/year. Thus annual net radiation values over equatorial South-East Asia and the Sahara desert are very similar. Mean annual temperatures are also similar over many of the land masses, and thus average annual sea-level temperatures in Indonesia are about 26°C, a value which is almost the same as the mean annual surface temperature over much of the Sahara.

During winter over the northern hemisphere, the net radiation values are low and become negative north of 40°N. They increase southward reaching values of 8 kilo-calories/cm^2/month over the equator and maximum values over the ocean near the Tropic of Capricorn. Over land in the southern hemisphere, net radiation values are relatively stable at about 6—8 kilo-calories/cm^2/month. In the northern summer the zero net radiation line lies about 45°S, north of which net radiation increases until it reaches maximum values over the ocean in the region of the Tropic of Cancer, with peak values of 12 kilo-calories/cm^2/month over the north Arabian Sea. Over land in the northern hemisphere during summer net radiation values are relatively constant with values fluctuating between 6—8 kilo-calories/cm^2/month. Thus values change little throughout the year near the equator, but the seasonal variations increase with increasing latitude.

Variations in the radiation climate are reflected by variations in seasonal temperatures. Thus in equatorial South-East Asia, the thermal regime is characterized by general uniformity, the sea-level temperature being almost constant throughout the whole area and throughout the year. At any given site the largest temperature variations are between day and night, rather than between seasons. At Singapore the annual temperature range is only about 2·2°C, but the diurnal range reaches 6·2°C. Over land areas both the diurnal and annual ranges increase with increasing latitude.

In the desert areas most of the net radiation is available for heating the air. At Aswan the average annual rainfall is below 2 mm, so the soil is nearly always dry, and summer afternoon temperatures regularly exceed 38°C and temperatures above 48°C have been recorded on a number of occasions. The hotter parts of the Sahara experience mean maximums of 45°C in the hottest month, and values nearly as high are probable in the Great Sandy Desert in western Australia, while Death Valley, California has a mean maximum temperature of 47°C in July. Seasonal variations in net radiation are large in desert areas, leading to correspondingly large seasonal temperature variations. The high summer temperatures at Aswan have been mentioned, but in January and February the night temperatures can fall as low as 5°C. Indeed, large areas of the Sahara have recorded absolute minimum temperatures of 0°C or below, indicating that frost is not unknown. At Alice Springs, in the desert of central Australia, night temperatures regularly fall below 0°C in June, July and August,

and the absolute minimum temperature is about −7°C.

Areas with near freezing average temperatures are limited in the tropics to the tops of high mountains. Over South-East Asia the mean altitude of the 0°C isotherm is 4 694 m, being highest in May at 4 846 m and lowest in September at 4 572 m. The snow-line, that is the line of permanent snow, is found therefore only on the most lofty peaks, which are in New Guinea, where on Sukarno Peak (5 030 m) it has been located at about 4 300 m.

While average temperatures below 0°C occur only over very limited areas in the equatorial zone, frost can be important over rather wider areas. Within the tropics, frost is usually due to intense radiational cooling rather than to the movement of cold air masses, and because of this its incidence is related to air temperature and humidity. Frost occurs in tropical mountain areas on clear, calm nights, in hollows into which cold air drains. In Java frost hollows are frequently found above 2 000 m, but although their number at about 1 500 m is restricted, nevertheless one even appears to exist at 900 m on the north side of the Idjen Caldera.

Frost is most frequent in the months of July, August, September and October when the atmosphere is often particularly dry over Java, in contrast to areas north of the equator where it is frequently very moist. The type of radiating surface is particularly important, since temperatures of −5°C to −10°C have been measured over grass at 2 000 m in Java, but the situation is entirely different if a similar site is covered by evergreen forest. The radiating leaf surfaces in evergreen forest are high above the ground, and the cold air sinks from these surfaces and mixes with warmer air below, and so a cold surface layer is not generated. Once the evergreen forest is cleared, the radiating surface is reduced to ground level and frost can be observed in areas which are previously frost-free. The mountains of Peninsular Malaysia are somewhat lower than those of Java, and as far as is known, the surface temperature has never fallen to freezing-point. The lowest minimum temperature observed in a thermometer screen at the Cameron Highlands, Peninsular Malaysia (1 448 m) is 2·2°C, which would suggest that on this occasion the temperature just above soil or short grass would be below freezing and a ground frost would occur. Since in Peninsular Malaysia the mountainous terrain reaches 2 100 m, it would appear that ground frost probably does occur in high enclosed hollows which are not covered by forest, and similar comments probably apply to other mountainous regions in South-East Asia.

FURTHER READING FOR CHAPTERS 3 AND 4.

Crowe, P.R. (1971). *Concepts in Climatology* (Longman, London).

Flohn, H. (1969). *Climate and Weather* (Weidenfeld and Nicolson, London).

Lockwood, J.G. (1974). *World Climatology: An Environmental Approach* (Edward Arnold, London).

Lowry, W.P. (1969). *Weather and Life: An Introduction to Biometeorology* (Academic Press, New York).

Riehl, H. (1965). *Introduction to the Atmosphere* (McGraw-Hill, New York).

Sellers, W.D. (1965). *Physical Climatology* (University of Chicago Press, Chicago).

PART C
Motion and Motion Systems

Typhoon damage in Hong Kong. The photograph was taken while the signposting on the right was collapsing; seconds later it was carried away by the wind. Reproduced by permission of the Controller of Her Majesty's Stationery Office; Crown Copyright reserved.

5 Motion in the Natural Environment

WITHIN the natural environment there are many forms of movement, some of which are obvious like river-flow or the winds, while others such as soil creep are less so. Since the same laws apply to all types of motion, it is necessary to look at their nature.

Physical quantities can be divided into scalars and vectors. A scalar quantity is one that has magnitude only, while quantities that have direction as well as magnitude are called vectors. Thus temperature, heat content, mass and frequency can all be specified completely by a number and are scalars, but velocity, acceleration and force are all vectors. While energy, which was considered in the previous section, was mostly concerned with scalars, motion involves mainly vector quantities.

At the end of the seventeenth century, Sir Isaac Newton formulated the basic laws that govern the motion of all matter, be it solid, liquid or gas. Newton's first law concerns the motion of matter that is not subject to any external forces. This law states that an object, if at rest, will remain at rest, but that if it is in motion, it will continue to move in a straight line. If a force acts on the object, then its motion will change according to Newton's second law, which states that a force will compel an object to accelerate (change its velocity) so that the magnitude of its acceleration is equal to the external force divided by the mass of the object. These two laws introduce the concept of a force, which is usually defined as any action which alters or tends to alter a body's state of rest or of uniform motion in a straight line. A commonly observed force is gravity, which causes objects to move towards the earth's surface.

A rigid body is one whose size and shape are not affected by the forces acting on it. The forces may create tensions within the body but they will not alter it. In contrast fluids offer very little or no resistance to change of shape when a force is applied.

There are two types of fluids—liquids and gases. A liquid is a fluid that cannot be compressed at all or that can be compressed only by applying very great pressure. It thus occupies a definite volume, and if placed in a container with a volume greater than that of the fluid, will have a free surface. A gas is a fluid of unlimited capacity for expansion under diminishing pressure, and will thus completely fill any container in which it is placed. Tensions may exist in solids for considerable periods of time without any changes occurring, but forces within a fluid must be the same in all directions, or otherwise the fluid will flow until the forces are equalized. Pressure in a fluid is therefore a scalar quantity and to describe the forces acting in a liquid it is only necessary to know the distribution of the scalar pressure, while the state of stress in a solid depends upon the direction as well as the magnitude of the forces acting.

Two forces control either directly or indirectly most of the motions found in the natural environment. Gravity is finally responsible for generating most motions and accelerations, while friction is largely responsible for keeping objects at rest or reducing velocities. The difference between the forces is that gravity accelerates objects vertically downwards towards the centre of the earth, while friction is a mechanical force of resistance which acts when there is relative motion between two bodies in contact. Soil and loose rock material on a steep slope experiences a force due to gravity which accelerates it down the slope, and it only remains in position because equally powerful frictional forces, which stop movement, are acting between the soil and rock particles. If the frictional forces are just slightly less than the gravitational force, then very slow movement may take place down the slope in the form of soil creep. Sometimes the frictional forces may suddenly decrease, due perhaps to the soil becoming saturated with water, and then the whole mass will move rapidly downhill as a landslide.

Gravity can also be responsible for producing motions within a fluid. The pressure intensity in a fluid at points on a level plane is less than at points at a lower level, since there is more fluid above the latter. This increase of pressure (Δp) with increasing depth (Δz) may be expressed in the form:

$$\frac{\Delta p}{\Delta z} = g\rho$$

where $\dfrac{\Delta p}{\Delta z}$ is the change of pressure with depth;

          g is the downward acceleration due to gravity;
and ρ is the density of the fluid.

In an undisturbed fluid, pressure will be equal in all horizontal surfaces, since no unbalanced forces are acting within the fluid. If pressure varies in a horizontal plane (note that it always varies in the vertical), the forces in the horizontal plane will not balance and fluid will flow from high pressure to low. This movement is due to the horizontal pressure gradient, the original spot pressures being created by the force of gravity. Variations of pressure within a horizontal plane could be induced by heat flowing into one part of the fluid and so generating a rise in temperature. Density varies with temperature, so a rise in temperature will cause a decrease in density and thus a fall in pressure. Many motions within both the atmosphere and oceans are generated by differential heating which results in unequal horizontal pressure distributions, the fluid then flowing under the influence of gravity.

Friction

Friction may be defined as the mechanical force of resistance which acts when there is relative motion of two bodies in contact, of a body in contact with a fluid, of adjacent layers of a fluid, or of adjacent fluids. Friction may be considered as a force which acts in the opposite direction to that in which motion is either taking place or tending to take place.

Some of the laws which govern the action of the frictional force may be explored by means of a simple experiment. Place a block of wood A of known weight on a table and attach to it a piece of string passing over a pulley and carrying a light pan B at the

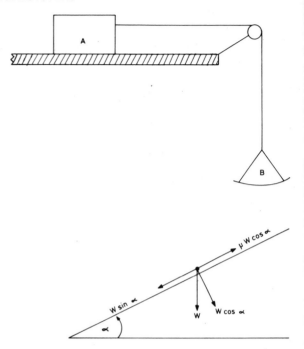

Figure 5.1 *Friction*
   Upper. A simple experiment to explore the laws of friction.
   Lower. The equilibrium of a particle on a rough inclined plane. See text for explanation.

other end. On placing a small weight on the pan B it is observed that no motion is produced in A. Thus the friction between A and the table must be equal to the weight of the pan B and the added weight. It is observed that the weight of B can gradually be increased until at a certain point the block A begins to move. This indicates that as the force tending to move A increases from zero, so the force of friction increases from zero at the same rate until a certain maximum or limiting value is reached, and then motion takes place. Further experiments show that the limiting value of friction before motion takes place, depends on the weight of A and the nature of the surface of A and of the surface of the table. For similar surfaces it is found that when motion is just about to take place the ratio of the total weight of B to the total weight of A is very nearly constant, and this is called the Coefficient of Friction (μ). When motion takes place, friction still opposes the motion but is slightly less than the limiting friction and is also independent of the magnitude of the velocity.

It is now possible to explore (Figure 5.1) the equilibrium of a particle on a rough inclined plane. Suppose a particle of weight W be placed on a rough plane whose inclination to the horizontal is gradually increased. At any inclination α, the component of the weight down the plane due to gravity is $W \sin \alpha$, while the pressure between the particle and the plane is $W \cos \alpha$. The limiting or maximum friction is $\mu W \cos \alpha$, so when $W \sin \alpha = \mu W \cos \alpha$ or $\tan \alpha = \mu$, motion is just about to take place. Thus the particle will begin to slide down the plane under its own weight when the angle of inclination is such that $\tan \alpha = \mu$, where in this particular case α is called the angle of friction. The above problem is similar to that of stones resting on a mountain side, where the stones will only remain at rest provided that the slope does not exceed some critical angle.

Viscosity is that property of a fluid whereby it resists deformation. Imagine (Figure 5.2) a fluid in which several different parallel layers move with different velocities, which increase in magnitude upwards. Friction within the fluid due to viscous forces produces stresses which tend to make the flow more uniform, and therefore the velocities of the upper layers are transferred downwards, causing the upper layers to slow down and the lower layers to quicken up. For two layers a short distance apart, both parallel to the direction of flow, the viscous stress per unit area (τ) is proportional to the velocity gradient, the constant of proportionality being known as the coefficient of dynamic viscosity (μ), i.e.

$$\tau = \mu \frac{\Delta u}{\Delta z}$$

where $\frac{\Delta u}{\Delta z}$ is the velocity gradient between the layers. The ratio of the dynamic viscosity to the density of the fluid is termed the kinematic viscosity.

So far it has been assumed that velocities are small and that viscosity arises mostly on the small molecular scale. In the atmosphere, turbulence occurs on a

very large scale causing mixing which evens out velocities in exactly the same way as molecular-scale viscosity. It is therefore possible to define, by analogy with the molecular kinematic viscosity, an eddy viscosity which results from the action of turbulent eddies. In air near the earth's surface, the kinematic viscosity has the approximate value of $1 \cdot 5 \ 10^{-5} \ m^2/s$ while the eddy viscosity is about $1 \ m^2/s$, that is 100 000 times greater.

The Earth's Rotation

Many of the large-scale motions upon the earth's surface are influenced by the rotation of the earth. This influence can be illustrated by exploring the motion of a particle on a flat rotating disc, which is shown in Figure 5.3.

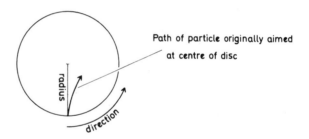

Figure 5.3 Motion on a Rotating Disc. Projectile moving towards the centre of the rotating disc appears to be deflected towards the right.

Since the disc is formed of a rigid solid, all points on the disc rotate around the axis in the same time interval. Now some points are further away from the central axis than others and therefore have to travel a greater distance during the period of rotation than points nearer to the axis. Because of the increasing distance to be travelled at greater distances from the axis, it follows that the actual speed of movement will increase with increasing remoteness from the axis. Thus the actual speed of movement will be greatest for points on the rim of the disc and least for points near the centre.

Consider a particle which is situated on the edge of a non-rotating disc. According to Newton's first law, an object not subject to external forces will move in a straight line at a uniform speed. So in the absence of friction, a particle projected from the rim towards the

Figure 5.2 Dynamic Viscosity. See text for explanation of symbols.

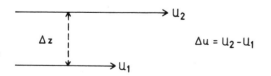

central axis of the disc will move along a straight line which will form one of the radii of the disc.

Imagine that the disc is rotating and that the particle moves across the disc in the absence of friction. At the starting-point on the rim, the particle is at rest relative to the rotating disc and it will have a velocity which is appropriate for its distance from the axis. If the particle is now projected towards the centre of the disc, its velocity will be made up of two components—the component towards the centre and a component at a right angle to the first due to the rotation of the disc. As the particle moves towards the centre it passes over parts of the disc where the speed of motion is less than that on the rim, so the particle, which retains its original speed, will appear to move sideways relative to the disc. If the disc rotates in an anticlockwise direction, the particle will appear to be deflected towards the right of the radius line from its starting position on the rim.

The earth's rotation about its axis at any point can be divided into three components (Figure 5.4), the most important of which is the one in a horizontal plane about an axis vertically upwards. At the poles

the horizontal component (about a vertical axis) will be equal to the rate of the earth's rotation, while away from the pole it will vary with sine (latitude), becoming zero at the equator. The horizontal component of the earth's rotation is such that the earth simulates a flat disc rotating anticlockwise in the northern hemisphere and clockwise in the southern. From the previous discussion it follows that particles moving horizontally outwards from the centre of the disc appear to an observer stationed there, to be deflected to the right in the northern hemisphere and to the left in the southern. To the observer on the rotating earth it appears that a force is acting on the particles which causes them to be deflected, and this apparent force per unit mass is termed the 'Coriolis force', or 'deviating force' or 'geostrophic force'. The Coriolis force is strictly a three-dimensional vector which is everywhere at right angles to both the earth's axis (in the plane of the equator) and the velocity of the object in motion, but normally only the horizontal component is of interest. The acceleration produced by the horizontal component of the Coriolis force is $2\,\Omega\,V\,\mathrm{Sin}\,\phi$, where Ω is the rate of the

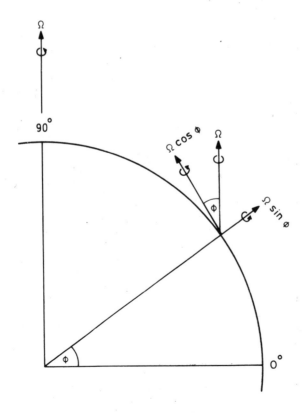

Figure 5.4 Horizontal and vertical components of the earth's rotation about its axis.
    Ω is the rate of the earth's rotation
    ϕ is the latitude.

earth's rotation, V is the velocity of the particle and ϕ is the latitude. This acceleration is always small, having a magnitude of $1 \cdot 5 \times 10^{-4} \times$ velocity cm sec^{-2} at the poles and decreasing to zero at the equator (Sin $(0°) = 0$). The quantity $2 \, \Omega$ Sin ϕ is known as the Coriolis parameter.

On the laboratory scale the Coriolis effect is masked completely by frictional forces, and therefore need not be considered. Similar considerations apply to small-scale movements in the natural environment, where friction effects are large and therefore rivers are not usually influenced by the Coriolis effect. Both the frictional and driving forces observed in large-scale motions in the atmosphere are small and are of the same order of magnitude as, or smaller than, the Coriolis effect, and so in both these cases the Coriolis effect is of some importance. The motion will be dominated by whatever force has the greatest magnitude, and it is only in very large-scale motion systems in the atmosphere and oceans that the other forces are small enough to allow the Coriolis force to become dominant.

Boundary Layers

From observations of fluid flow through pipes, it is possible to distinguish between two types of motion, called laminar and turbulent. This may be done by introducing some dye into the middle of a liquid current flowing through a thin glass tube and observing its behaviour. When the flow velocity is low, a well-defined dye filament is visible, the fluid layers appearing to follow parallel courses; but when the flow is increased, the filament is torn apart and the dye distributed throughout the whole fluid, whose motion has now become very irregular. The first type of flow is termed laminar, while the second is called turbulent.

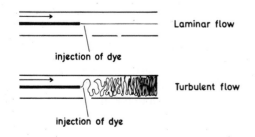

Figure 5.5 Laminar and turbulent flow.

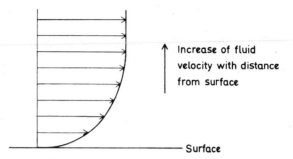

Figure 5.6 Boundary layer over a smooth surface. The fluid velocity increases above the smooth surface until it becomes essentially that of the mean flow in the main stream. The boundary layer is the layer where the fluid velocity is greatly influenced by the fixed surface.

In laminar flow each particle follows the precise path of its predecessors, and obviously an essential feature of such flow is that there is no mixing of adjacent layers of the fluid. Turbulent motion contains random oscillations which lead to irregularities in the paths of individual particles. While there is no precise mathematical definition of turbulence, it is generally taken to comprise the complex spectrum of fluctuating motion which is superimposed on a 'mean flow'. A mean flow cannot be defined without specifying the time over which the average flow is defined, and thus the mean flow may be an average for a minute, a day, or a year, and fluctuations about the mean are called turbulence. It is therefore obvious that there can be some debate as to what constitutes turbulence and what constitutes mean flow, since this will depend on the time period over which the flow is averaged. A fundamental property of turbulence is the vertical interchange of mass, sensible heat and water-vapour, and this is effected by eddies of a variety of shapes and sizes. Turbulence is particularly important because the effect of eddies is to increase mixing in fluids far beyond the rates which would be appropriate to purely molecular action.

When a fluid flows over a smooth solid surface (Figure 5.6), the fluid layer in contact with the surface will be at rest because of the friction between the fluid and the plate. Above this layer the fluid velocity increases slowly until at some level it is essentially that of the mean flow in the main stream. This layer, where the velocity of the fluid is greatly influenced by the fixed surface, is known as a boundary layer.

In a very shallow layer of air, adjacent to the fixed boundary, fluid velocities are very low and the flow is laminar, there being no turbulence and the viscosity resulting from molecular scale stresses. This layer is known as the laminar boundary layer, and in the atmosphere it is typically about 1 mm deep over a smooth surface. Since there is no turbulence in the laminar layer, heat can only cross it by the process of conduction, while mixing can only take place by the process of diffusion. Conduction may be defined as the process of heat transfer through matter by molecular impact from regions of high temperature to regions of low temperature without the transfer of the matter itself. In contrast, molecular diffusion is the process by which contiguous fluids mix slowly, despite differences in their density. The process follows similar laws to those of thermal conduction, that is to say, diffusion of a substance (e.g. water-vapour) takes place from regions of high density towards those of low density at a rate proportional to the density gradient. Both conduction and diffusion are very slow processes in the atmosphere, and are unimportant compared with mixing effected by convection and turbulent eddies.

Heat and water-vapour are transported through the laminar boundary layer by conduction and diffusion respectively, but as turbulence increases with distance from the ground, eddies gradually become the more important transporting agents. Thus over hot damp ground eddies carry warm humid air aloft, and this is replaced by cool dry air which in turn has its temperature and humidity increased. Within about the first metre of the atmosphere the rate at which water-vapour and sensible heat are carried upwards will on average be the same at each level. The good insulating properties of the laminar boundary layer cause very steep gradients of both temperature and humidity to exist within it, for otherwise the flow of heat and water-vapour will be less than that at higher levels. On a clear hot day in the tropics temperature gradients of up to 30°C per mm can exist in the laminar boundary layer.

The influence of the earth's surface on air flow is often detectable up to an altitude of 2 km, and this layer is known as the planetary boundary layer. It can be divided (Figure 5.7) into several sub-layers, the lowest of which is the laminar boundary layer. Other layers, above the laminar boundary layer, are the surface or logarithmic layer and the spiral layer.

Within about 20 m of the surface, there exists a layer in which the wind usually increases with height

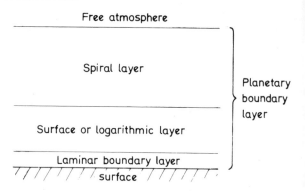

Figure 5.7 The Sub-layers of the planetary boundary layer. The depth of the planetary boundary layer varies between about 1 and 2 km.

in a logarithmic manner. This is illustrated in Figure 5.8 where wind speed is plotted against the logarithm of the height, since it is observed that the values fall on a straight line. When several series of observations are taken at the same site, it is found that while the slope of the lines drawn through the individual sets of observations vary, the intercepts of the lines with the height axis is always about the same value. This suggests that the wind speed becomes zero at some particular height above the ground, and that this height is nearly independent of the wind velocity aloft. The magnitude of the height (Z_0) at which the wind speed becomes zero is called the roughness length and it is found to be mainly a function of the type of surface. Thus for smooth surfaces it is small and it increases in magnitude with the increasing roughness of the surface. Typical values are 0·01 to 0·1 cm for sand, 0·1 to 0·5 cm for snow surfaces, 0·5 to 1·0 cm for mown grass and 4 to 10 cm for long grass.

Above the logarithmic layer the wind increases more slowly until at the top of the spiral layer it becomes equal to that in the main bulk of the lower atmosphere. In the spiral layer not only does wind speed change with height but so also does the wind direction, and hence the name of this layer. At the top of the spiral layer the wind can to a good approximation be considered to be equal to the geostrophic wind, which is an imaginary wind blowing parallel to the isobars (lines of equal pressure) at a speed which is inversely proportional to the spacing of the isobars. On a non-rotating earth the winds would blow directly down the pressure gradient, but on a rotating earth (Figure 5.9) the wind velocity assumes a magnitude

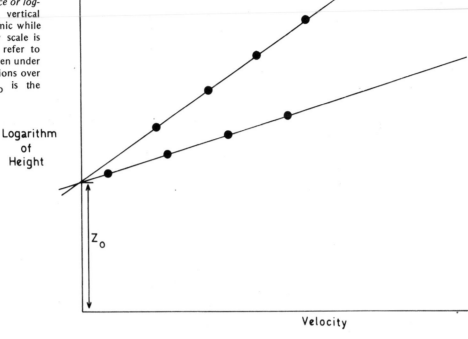

Figure 5.8 Variation of wind with height in the surface or logarithmic layer. The vertical height scale is logarithmic while the horizontal velocity scale is linear. Dots and lines refer to sets of observations taken under neutral stability conditions over the same surface. Z_0 is the roughness length.

Logarithm
of
Height

Z_0

Velocity

so that the horizontal component of the Coriolis force balances the pressure gradient force. That is:

horizontal pressure gradient force = $V_G f \rho$
where V_G is the geostrophic wind;

where f is the horizontal component of the Coriolis force

$(2 \Omega \sin \phi)$;
and ρ is the air density.

Figure 5.9 The geostrophic wind.
   V_G is the geostrophic wind.
   f is the horizontal component of the Coriolis force.
The pressure gradient force acts from high to low pressure.

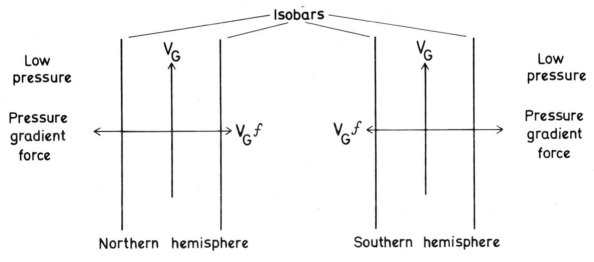

Isobars

V_G V_G

Low
pressure

Low
pressure

Pressure
gradient
force

$V_G f$ $V_G f$

Pressure
gradient
force

Northern hemisphere   Southern hemisphere

Thus low pressure is to the left of the wind arrow in the northern hemisphere, and to the right in the southern. Below the top of the friction layer the wind is increasingly deflected towards low pressure, until near the surface there is marked cross-isobaric flow and the wind speed is considerably less than the geostrophic value.

The Circulation of the Atmosphere

The planet earth receives heat from the sun in the form of short-wave radiation, but it also radiates an equal amount of heat to space in the form of long-wave radiation. This balance of heat gained equalling heat lost only applies to the planet as a whole over several annual periods; it does not apply to any specific area or for a short period of time. In particular the equatorial region absorbs more heat than it loses, while the polar regions radiate more heat than they receive from the sun. These inequalities in surface heating create horizontal density gradients in the atmosphere, which in turn under the influence of gravity cause atmospheric motions. Thus the circulation of the atmosphere is driven by differences in surface heating, and it will be shown later that it is partly shaped by the earth's rotation.

Over an annual period, most global radiation is received by the earth's surface between 35°N and 35°S, while there is a gradual decrease towards the polar regions. On the absolute temperature scale the horizontal variations in the temperature of the earth's surface and lower atmosphere are small, and so the variations in the long-wave radiation loss between the equator and the poles are small in comparison with those in global radiation. Thus a situation is created (see Figure 5.10) where on the annual scale the net radiation is positive between about 40°N and 40°S and negative at higher latitudes.

The radiation balance of the earth is positive during the whole year only in the narrow equatorial zone between the latitudes 10°N to 10°S, for elsewhere the sign of the net radiation changes twice a year. For about three summer months in each year the radiation balance of the whole of each hemisphere is positive, but in late summer zones of negative radiation arise near the poles and then gradually spread toward the equator, reaching about latitude thirty after five months, while a similar process of retreat occurs in the spring.

Although there is a long-term global balance between incoming and outgoing radiation, considerable imbalances exist both locally and seasonally, and it

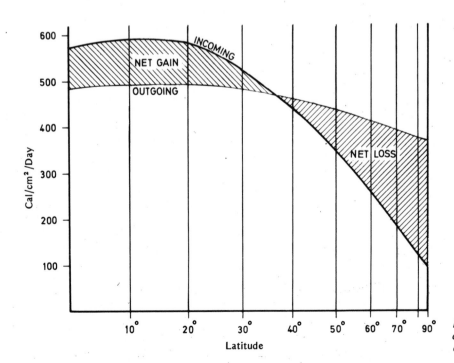

Figure 5.10 Annual latitudinal distribution of incoming and outgoing radiation.

has already been shown that there is a substantial excess of net radiation in low latitudes and a deficit towards the poles. Alternatively, it can be considered that the mean atmospheric temperature in equatorial latitudes is lower, and in polar latitudes higher, than those appropriate to the local radiative balance, and that this situation is possible because of the global-scale mixing performed by the general atmospheric circulation. The atmospheric circulation is maintained against the various frictional forces which tend to destroy the motions by the heat energy from the sun, and it would assume a very simple form but for the modifying influence of the earth's rotation. This simple circulation, in the absence of the earth's rotation, would probably consist of two simple cells with rising air over the equator and sinking air over the two poles.

The earth's rotation complicates the general circulation in a number of ways, some of which are not as yet fully understood. The angular momentum per unit mass of a body rotating about a fixed axis is the product of the linear velocity of the body and the perpendicular distance of the body from the axis of rotation. Now in the upper atmosphere, in the absence of significant friction or other forces, angular

momentum remains constant over periods of a few days. Rings of air flowing poleward at high levels from near the equator move nearer to the earth's axis, because of the spherical shape of the earth, and since the angular momentum is conserved, the eastward velocity of the air increases. Thus, on average, the eastward velocity of the air between about 6 km and 14 km increases away from the equator, reaching velocities of 30 to 60 m/sec between 30° and 40° N and S, appearing on weather-charts as the sub-tropical westerly jet streams.

A jet stream is a fast narrow current of air, usually found in the upper troposphere, and it is generally some thousands of kilometres in length, hundreds of kilometres in width and a few kilometres in depth. The sub-tropical jet streams (Figure 5.11) mark the poleward limit of the outward flow from the equator, and there is extensive subsidence out of them. Now high pressure at the surface is usually associated with sinking air, and low pressure with rising air, so zones of high pressure (Figures 5.12 and 5.13) exist in the sub-tropics below the sub-tropical jet streams. At the surface, air flows out of the sub-tropical anticyclones towards the equator and the middle latitudes, the equatorial return flow forming the north-east and

Figure 5.11 Schematic representation of the meridional circulation and associated jet stream cores in winter.
  STJ subtropical jet stream
  PFJ polar front jet stream

(After E. Palmen (1951). 'The role of atmospheric disturbances in the general circulation', *Quart. J. Roy. Meteorol. Soc.* 77, p. 337).

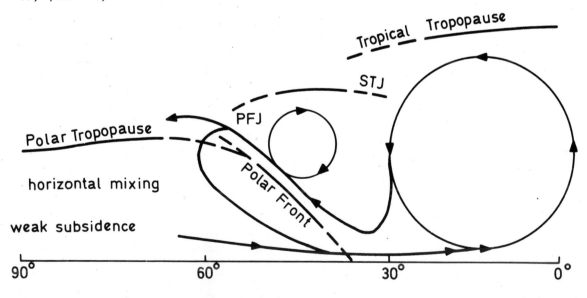

Latitude

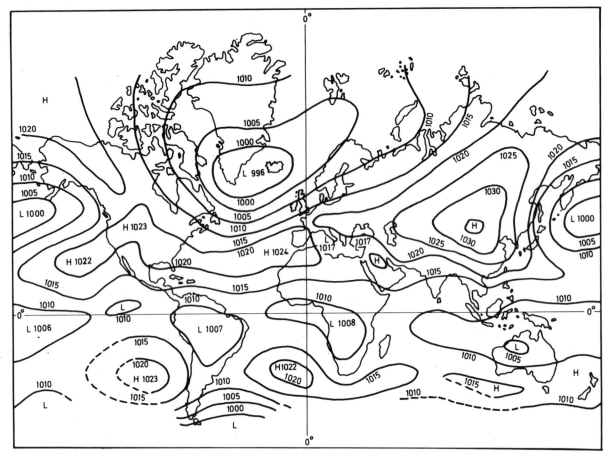

Figure 5.12 Average atmospheric pressure (mb) at sea level in January.

south-east trade winds. Thus the rotation of the earth limits the simple direct circulation cells (known as the Hadley cells) to the tropics.

In the tropics the horizontal component of the Coriolis force is small, but in middle latitudes it is much larger and calculations indicate that under these conditions the general flow will be towards the east and that it will be highly disturbed, containing waves in the upper atmosphere and trains of moving vortices near the surface. The flow in middle latitudes is like this, containing many waves and jet streams in the upper troposphere and series of travelling highs and lows at the surface.

Over the poles the radiational cooling of the atmosphere causes weak subsidence and weak outflowing easterlies which feed into the travelling vortices of the temperate latitudes. The average north-south flow of the atmosphere can therefore be considered to

consist of three circulation cells, as shown in Figure 5.11, with easterly winds with a large equatorial component in the tropics, high pressure in the sub-tropics, westerly winds with a poleward component in middle latitudes and weak easterlies near the poles.

Any small region of the atmosphere contains two components of rotation about a vertical axis. The first is the horizontal component of the earth's rotation and the second is due to local circulations within the atmosphere. Since the earth rotates towards the east, in an anticlockwise manner in the northern hemisphere, the first component is also anticlockwise in the northern hemisphere. In a low pressure system there is general convergence of air at low levels towards the centre of the system, and the air imported contains the dominant anticlockwise spin due to the earth's rotation. The convergence into the system intensifies the latent rotation of the air and produces

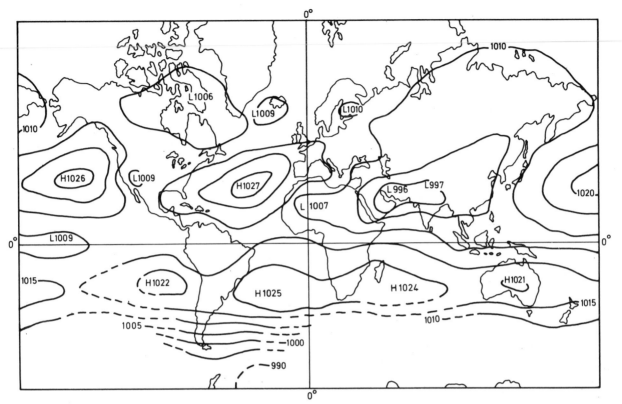

Figure 5.13 Average atmospheric pressure (mb) at sea level in July. (Both Figures 5.12 and 5.13 after H.H. Lamb (1969).)

Climatic fluctuations. In H. Flohn, (Ed). *General Climatology 2.* (Elsevier, Amsterdam).)

the cyclonic spin of the depression. In the southern hemisphere the horizontal component of the earth's rotation acts in the opposite sense, so clockwise rotation occurs in low pressure systems. General outflow takes place at the surface in anticyclones (high pressure systems) and this leads to the opposite effect to that observed in depressions, that is clockwise rotation in the northern hemisphere and anticlockwise in the southern.

An air mass is a body of air in which the horizontal gradients of temperature and humidity are relatively slight and which is separated from an adjacent body of air by a sharply defined transition zone, known as a front, in which these gradients are relatively large. Horizontally homogeneous bodies of air are produced by prolonged contact with an underlying surface of uniform temperature, known as a source region. Source regions must have light winds and are therefore usually found in the permanent or semi-permanent high pressure systems—the subtropical, polar and winter continental anticyclones.

This leads to a general classification of air masses as 'polar' or 'tropical', maritime or continental, defining their basic temperature and humidity characteristics.

Tropical air originating over the oceans in the sub-tropical highs around 30°–35° N and S is known as maritime tropical air. It is quite warm and, near the surface, moist; it forms the trade winds and also flows into the westerlies of the temperate latitudes. Continental tropical air forms over the large deserts of the tropics, and it is characteristically extremely hot and dry at the surface, but when it flows over the sea it is gradually modified to form maritime tropical air.

Polar air originates in high latitudes and may be subdivided into maritime polar and continental polar according to the nature of the surface over which it formed. The former is only relatively cool and usually very moist and it feeds into the poleward edge of temperate latitude westerlies. Extensive anticyclones form in winter over the continental interiors of Asia and North America and these are the sources of extremely cold, dry winds which blow out of the continental interiors in winter.

The general circulation of the atmosphere produces a broad latitudinal zonation of the world climatic zones. Thus a warm humid zone is found near the equator corresponding to the ascending limb of the Hadley cells, but in contrast the descending limbs in the sub-tropics cause widespread aridity. The travelling disturbances of the middle latitudes create cool, humid zones poleward of the sub-tropical deserts, while precipitation is low in the polar regions because the air is too cold to hold much water-vapour. This very simple climatic zonation is distorted by the continents. The precipitation-forming disturbances of the temperate latitudes rarely penetrate into the continental interiors, which therefore remain dry and tend to be poleward extensions of the sub-tropical deserts. The largest modification is found over southern Asia where the normal wind circulation is reversed during the northern summer, bringing rain to areas which would normally be sub-tropical desert.

Thus the thermally driven motions of the atmosphere are basically responsible for creating the various climatic zones, and therefore form an important part of the natural environment.

Depressions and Anticyclones

Polar and tropical air masses meet in middle latitudes along a front which is given the general name 'polar front'. The warm tropical air rises over the cold polar air, so the front rises towards the pole with a slope of about 1 in 200. The polar front is usually undisturbed, being marked by a line of cloud and a little light rain, but occasionally a small wave forms on the front, the warm air penetrating horizontally into the cold air. Some of these waves grow rapidly in a period of one or two days, as shown in Figure 5.14, and in particular the penetration of the warm air in the form of the warm sector increases, resulting in a very pronounced wave structure. Much of the ascent and weather in a frontal depression takes place along the warm and cold fronts. At the warm front, warm air displaces cold along a gently sloping surface of about 1 in 200, spreading rain and layer cloud over a wide area ahead of the depression, while at the cold front cold air displaces warm air along a steeply sloping surface of about 1 in 50, generating a narrow belt of cloud and rain.

The initial wave disturbance grows in a period of about a day into the young frontal depression shown in the diagram. As the warm air ascends and escapes to higher levels over the warm-front surface and as the cold front undercuts the warm air, the warm sector narrows, with the result that the cold front tends to overtake the warm front, and the cyclone is said to be occluded. The front resulting from the combination of the warm and cold fronts is called an occlusion. In the early stages of depression development, the lowest pressure is found at the tip of the wave formed in the polar front, but as the occlusion process continues, the fronts gradually become separated from the region of lowest pressure. In the later stages of the life cycle, the occlusion process becomes almost complete, and the warm air is lifted completely off the surface. A frequently-occurring feature of the partly-occluded and occluded stages is the development in the rear of the depression of a trough, which is often accompanied by bad weather due to general convergence in the lower troposphere.

In the northern hemisphere, depressions frequently form in winter off the eastern coasts of Asia and North America. During the early stages of development, the cyclones move rapidly north-east, but after 24–30 hours they start to occlude and slow down. They frequently become stationary over the eastern parts of the oceans, and slowly decay over several days. Indeed, it is the frequency with which depressions become stationary near Iceland or the Aleutian Islands which gives rise to the Icelandic and Aleutian lows on mean pressure charts (see Figures 5.12 and 5.13). Furthermore, it is observed that frontal depressions rarely appear alone, for normally there are two, three or more in a series, each in the wake of the other with a general tendency to move north-east.

Frontal depressions are primarily maritime storms which are best developed over the oceans. In the interior of continents, particularly to the east of the Rocky Mountains and in central and eastern Eurasia, the areas of continuous cloud cover and precipitation may be small, and in many cases the precipitation area may be broken or absent. Indeed, the weather distribution within a cyclone depends very much on the location and the time of year, and it is not possible to generalize with safety. For example, cold fronts are often very active with severe thunderstorms over the eastern U.S.A., but they rarely give any significant weather in western Europe.

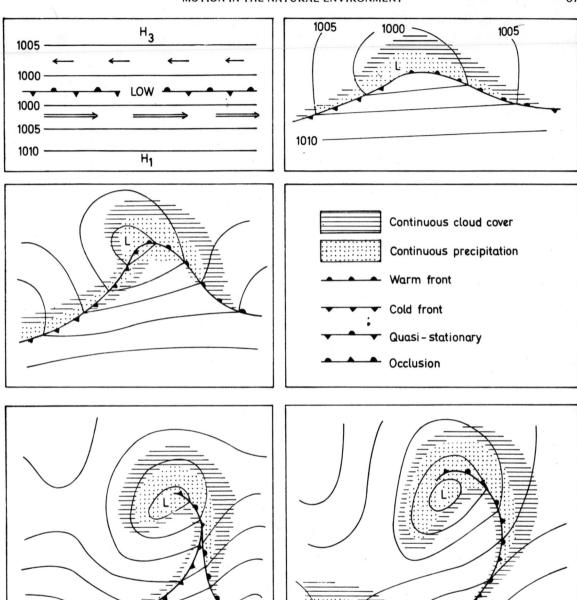

Figure 5.14 The development of a frontal depression. This is the northern hemisphere case with cold air to the north and warm air to the south. The initial undisturbed polar front is shown in the upper left, while the occluded state is in the lower right. Winds blow approximately along the isobars, which are shown by thin continuous lines. (After S. Petterssen (1956). *Weather Analysis and Forecasting* (McGraw-Hill, New York). Used with permission of McGraw-Hill Book Company).

Anticyclones are the other major weather system found in temperate latitudes. They often appear as sluggish and passive systems which fill the spaces between the far more active depressions. Anticyclones may be divided very broadly into cold or polar continental highs and warm or dynamic highs.

Polar continental highs develop in winter over the northern land masses, where they are caused by the intense cooling of the snow-covered surface, giving rise to a shallow and dense layer of very cold air. The relatively high density of the cold air increases the surface pressure above normal and so an anticyclone appears on the surface pressure chart. These cold anticyclones are therefore only shallow formations and it is rare that they can be traced more than 2 km above sea-level. The most pronounced and persistent example (Figure 5.12) is found over Siberia, the corresponding high over North America being far less regular. Both these highs are, of course, the main sources of continental polar air in the northern winter.

The warm or dynamic anticyclone (Figure 5.15) is caused by large-scale subsidence throughout the depth of the lower atmosphere, and good examples are the highs associated with the descending limbs of the Hadley cells. An essential feature of all dynamic anticyclones is a low-level temperature inversion. The general subsidence usually ceases at the top of this inversion and the weather associated with the high often depends on the nature of the comparatively cool air near the surface. If the surface air is moist, cloud may form below the inversion and light rain may even fall, and this is particularly likely to occur where the flow is from a warm sea across a cold land or sea surface. In contrast, when the surface air is relatively dry, the skies may be almost cloudless, leading to great heat in summer and intense cold in winter. As with depressions, it is not possible to generalize about the weather distribution within these systems. Semi-permanent dynamic anticyclones are common in the sub-tropics, but they can also occur anywhere in the middle latitudes at any season of the year.

The Wind-driven Circulation of the Oceans

Winds blowing over the oceans produce waves and cause the surface water to drift in the direction of the air flow. Waves may be generated by dropping a stone into a calm pond, when it is observed that the waves travel outwards from the point of disturbance without any corresponding movement of the water. A cork placed on the surface moves up and down and does not travel with the waves. So waves on a water surface do not necessarily imply a forward movement

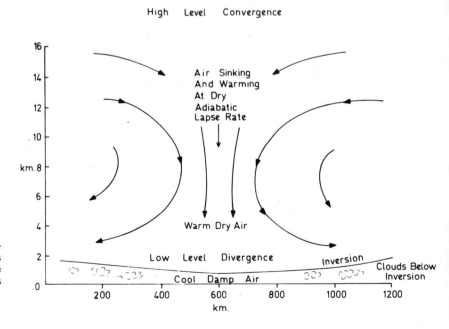

Figure 5.15 Schematic representation of vertical motions in a dynamic anticyclone. Superimposed on the vertical motions may be horizontal flows in the form of high-level jet streams and low-level rotations.

of the water, and after simple wave motion, each water particle returns to the point from which it started.

When the wind blows over a wave on a water surface the pressure on the windward side of the wave is higher than the pressure in the wind shadow of the wave. Thus there is a tendency to push the wave in the direction of the wind, and also as the wind moves over the water surface, it exerts a frictional drag which moves the surface water in the direction of the air flow. As a result of these factors, the water particles in the waves do not return to their original positions but are dragged slightly in the direction of the wind.

Ocean waves contain potential energy due to the vertical displacement of the water and also kinetic energy due to the particle motion. On average they are equal over a unit area of ocean surface and are proportional to the square of the wave height, hence the great destructive power of very large waves. Kinetic energy within the wave is gradually dissipated by internal friction and unless new energy is added, the wave will be reduced in height. The wind imparts energy to the water surface, so the wave height will eventually be such that the rate of loss of energy in the wave is equal to the rate of input of energy from the wind, that is to say, the higher the wind speed, the greater the ultimate wave height. The actual rate of dissipation of the wave away from its original source seems to depend on the wave-length (distance from crest to crest). Waves with short wave-lengths are observed to dissipate rapidly outside of the storm in which they were generated, while those with long wave-lengths travel virtually without attenuation halfway around the world before spending their energy on a distant beach.

If the earth were entirely covered by water, the circulation of the ocean surface would be similar to that of the sea-level winds in the atmosphere, because of the drag exerted by the winds. The oceans are separated into distinct basins (Figure 5.16) by the continents which deflect the wind-driven currents, resulting in a series of circular currents, or gyres. These are best observed in the two major oceans, the Atlantic and the Pacific, both of which extend from the Arctic to the Antarctic. The largest gyre, the subtropical gyre, is formed by water being dragged to the west by the trade winds and returning eastward in the westerlies. A small equatorial gyre is also formed by

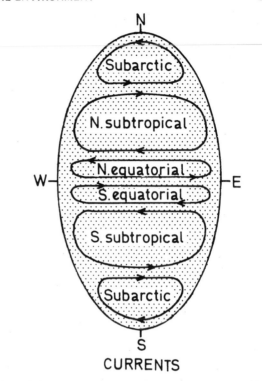

Figure 5.16 *The wind-driven circulation of the oceans.* The major gyres are indicated.

the partial return to the east, near to the equator, of water that has been displaced in the trade winds. Poleward of the sub-tropical gyre is a sub-polar gyre formed by the polar easterlies and temperate westerlies. In an ideal ocean there should therefore be three major gyres on each side of the equator.

Variations in solar radiation and mixing with water from adjacent gyres cause variations in the temperature of the ocean currents. In the sub-tropical gyre the westward flowing water near the equator is heated, both by the high insolation and also by mixing with water from the equatorial gyre, while the eastward flowing water near the pole is cooled, both because of the low insolation and also by mixing with water from the polar gyre. As a result, the water temperature of the poleward flowing limb in the west will be higher than that of the equatorward flowing limb in the east. Similar consideration applies to the other gyres, thus creating a series of warm and cold currents along the continental margins. For example, in the Pacific the poleward-flowing Kuro Shio and East Australian Currents are warm, while the equatorward-flowing California and Peru Currents are cool.

The sub-tropical gyres show considerable asymmetry, the centres being displaced towards the west, and therefore the poleward water flow is concentrated within a narrow band along the western sides of the oceans. In particular the Gulf Stream and Kuro Shio are vigorous currents as compared with currents observed elsewhere, and current velocities can reach 1–5 m/sec in narrow streaks which are similar to the jet streams observed in the upper troposphere. The asymmetry of the sub-tropical gyres is considered to be the result of the variation of the Coriolis parameter with latitude.

6 The Weather and Climate Systems of the Tropics

As is often the case in geography, it is not possible to draw boundaries that will exactly define the tropics. Some people say that the Tropics of Cancer and Capricorn mark the boundaries of the tropics, while others point to the 30th parallel, which divides the earth's surface between pole and equator into equal halves, but anyone living in the marginal belt will find both definitions arbitrary. Central China experiences tropical weather in summer but rarely in winter; for example Changsha (28° 15'N, 112° 20'E) in Hunan Province has a July mean temperature of 29°C but in January the mean temperature falls to 4°C. In the context of this chapter it is probably best to define the meteorological tropics as that part of the world where most of the time the weather sequences differ distinctly from those in middle latitudes. The axes of the sub-tropical anticyclones act as a rough guide to this particular boundary.

The tropics differ meteorologically from the middle latitudes in a number of important aspects. In the tropics the surface winds are normally easterly and temperatures tend to be uniform in the horizontal over vast areas, with the result that contrasting air currents are rare, whereas in middle latitudes winds are generally westerly and there are very marked north-south temperature gradients with air currents of different origins possessing markedly different temperatures and humidities. In the middle latitudes the Coriolis parameter is marked, and therefore there is a definite relationship between wind and pressure, but within about 5° of the equator the Coriolis parameter becomes very weak, marked cross-isobaric flow can occur and closed circulations are unusual except where they are topographically induced.

Tropical Air Masses

Before considering flow patterns within the tropics

TABLE 6.1

PROPERTIES OF AIR STREAMS IN THE VICINITY OF THE PHILIPPINES

Air stream	Month	Surface Temperature (°C)	Surface Relative Humidity (per cent)
South-west monsoon	Mar.–May	27·2	84
	June–Aug.	25·5	82
	Sept.–Nov.	25·5	82
North Pacific Trades	Dec.–Feb.	26·6	59
	Mar.–May	28·8	60
	June–Aug.	27·2	68
	Sept.–Nov.	27·0	74

it is necessary to look at the climatic characteristics of tropical air. A large number of names are used to describe tropical air masses, but there are only really two broad divisions—maritime tropical and continental tropical. Because of the large area of the tropics covered by ocean, maritime tropical is the most common air mass type, and will therefore be considered first.

Upper air observations on tropical islands show changes of temperature and moisture content from day to day, but the amplitude of these changes increases upward, in marked contrast to middle latitude areas where they decrease with increasing altitude. Near the surface over the tropical oceans, temperature variations seldom exceed 1° or 2°C, and these variations are entirely in the range of error of the measuring instruments and local modifications of the air due to cloudiness and evaporation from falling rain. It would seem therefore that a nearly uniform air mass covers vast areas of the tropical oceans. Many disturbances containing cloud and rain exist over the tropical oceans, and it appears that the air itself, moving with the winds passes from one side of these disturbances to the other in a few days. Therefore many of the tropical weather systems form within a uniform air mass. Tropical air encountered over the great oceanic regions does not possess a deep moist layer, for usually there is an inversion about 2 to 3 km above sea-level and the moist air is trapped below this inversion. Above the inversion the atmosphere is dry and normally cloudless. Table 6.1 illustrates some of the properties of typical maritime tropical air reaching the Philippines, and it can be seen that the variation in surface temperature is very small.

Extreme heating which takes place over the subtropical deserts leads to a second type of tropical air mass known as continental tropical air, of which the main sources are North Africa and the Middle East in summer. Its outstanding characteristic is its great heat often reaching 40° or 50°C at the surface. The air is often stable, inversions can be found at about 2 km and the relative humidities are usually extremely low. Typical air mass properties of tropical continental air at Khartoum are listed in Table 6.2.

In winter, areas on the borders of the tropics can be influenced by middle latitude air masses. A winter anticyclone is nearly always centred over Mongolia, where mean winter temperatures are below −10°C and often below −20°C. Continental polar air outflowing from this continental anticyclone spreads down to the North China Plain, and successive cold waves sweep south-eastward across China and Korea at intervals of about one week throughout the winter. These cold waves only extend 1·5 to 2 km above ground level, so their progress is greatly affected by topography. Thus the southern provinces of China are sheltered to some extent by the Nanling Hills, but

they are quite exposed to any cold outbursts which travel along the coast. This occurs when cold air from Mongolia has passed across the Yellow Sea to be funnelled south-westwards through the Taiwan Channel to Fukien and Kwangtung. These cold air masses are soon modified by warming over the tropical oceans, and therefore rarely penetrate far into the tropics in a completely unmodified form. Normally the warming by the tropical oceans is so effective that they very quickly assume the properties of tropical oceanic air. For instance, in winter, air leaves Mongolia as a continental polar air mass with a temperature of between −10°C and −20°C, and as it passes over the Pacific, is transformed into a maritime polar air mass. It finally reaches the Philippines as a maritime tropical air mass, with a surface temperature of about 25°C.

The Sub-tropical Anticyclones

Two belts of high pressure (Figures 5.12 and 5.13) at about 30°N and 30°S contain several quasi-permanent anticyclones known as the sub-tropical anticyclones, separated from each other by cols. The most notable ones in the northern hemisphere are the two oceanic highs, in the Pacific and the Atlantic respectively, and the North African high, which fails to show up at sea-level but emerges clearly at the 3-km level. In the southern hemisphere semi-permanent anticyclones are found over the Pacific, Atlantic and Indian Oceans.

The sub-tropical anticyclones which tend to be located at fixed positions on the globe, show great permanence and only slight seasonal variations. With regard to the geographical equator, the sub-tropical high-pressure ridges reflect a slight asymmetry, the southern ridge being situated 5° latitude closer to it in the mean than the northern one. They show only small seasonal displacements, amounting to about 5° latitude on average, as they are nearest to the equator in winter. At the sub-tropical ridge lines, pressure is practically equal in both hemispheres, varying on average from 1 015 mb in summer to 1 020 mb in winter.

Sub-tropical anticyclones are formed by air subsiding from high levels in the atmosphere, in this case out of the sub-tropical jet streams and, typically, the air may take about three weeks to subside from 12 km to 3 km. Because the air is sinking it is warming adiabatically, and therefore the anticyclones are

TABLE 6.2

PROPERTIES OF TROPICAL CONTINENTAL AIR
AT KHARTOUM

	Temperature (°C)	Relative Humidity (per cent)
Summer	41	16
Winter	32	24

cloud-free in the middle layers. An inversion (Figure 5.15) is always found in the bottom 2 or 3 km of a dynamic anticyclone, since the warm subsiding air does not sink right to the surface but spreads out over a layer of relatively cool air in which convection normally takes place and cumulus clouds form. The low-level inversion forming in the anticyclones becomes incorporated in the trade winds as the trade-wind inversion which is carried almost to the equator. If an anticyclone exists over a hot desert, such as the Sahara, the lower layers of the anticyclone may be distorted by the intense heat and a surface low will appear. This heat low is a shallow surface feature and the clear subsiding anticyclonic air will be found above it.

The continual subsidence in the sub-tropical anticyclones makes it almost impossible for extensive clouds to form and therefore for rain to fall. The important arid zones of the world are found around latitudes 30°N and 30°S, where large areas are dominated by anticyclones all the year round. Situated in these latitudes are most of the large deserts of the world including the Sahara in North Africa, the Arabian and Syrian deserts, Death Valley in North America, the Atacama Desert in South America, the Kalahari Desert in southern Africa and the Australian deserts. The great sub-tropical rainless zones spread over the oceans as well as the land masses for the lack of rainfall is not due to the absence of surface water.

Topography can increase the aridity observed in the sub-tropics. The general atmospheric subsidence may be intensified by the air sinking on the lee side of mountain ranges into enclosed inland basins, for the most severe desert conditions in the world are found under such circumstances, e.g. Death Valley in the sub-tropical regions of the western U.S.A. On the other hand, along the coast from the Atacama Desert of northern Chile to northern Peru, extreme aridity is caused by the combined effects of the circulation of the atmosphere and the cold Humboldt Current, which runs from the Antarctic along the west coast of South America. A cold sea current, flowing away from a coast, causes relatively cold bottom water to well up along the coast, adding an adiabatic cooling effect in the rising water to the original coldness. The unusually cold water acts as a strong stabilizing factor on the air above and greatly intensifies the normal low-level anticyclonic inversion. At the Atacama and Peruvian coasts, rain is practically non-existent due to

the stability of the air, since the trade-wind inversion is being strengthened, not only by subsidence higher up, but also by cooling from below. Because of this, the base of the inversion can be very low, and has been reported in similar circumstances as having descended to less than 50 m above the sea, but the average altitude of the base of the inversion appears to be about 500 m. The air below the inversion can be quite humid and fog is a common phenomenon, but the humidity of the air is not sufficient for the growth of trees or important vegetation in the absence of rain. It is obvious that when the effect of a cold ocean current is added to the other factors creating aridity, the consequences become extreme and this is true locally in many areas along the western coast of South America. Other regions where weaker cooling effects of the sea are operative occur along the Californian coast and on the west African coasts in both the northern and southern hemispheres.

No desert area in the world is completely rainless, for although in most arid areas of the sub-tropical anticyclones it may only rain once every few years, it does occur. When desert rainfall takes place, it can be extremely intense, and it is not uncommon for the whole annual rainfall to fall in one storm. If the storm is an extreme one, it may even equal several years' rainfall, and such heavy falls can give rise to floods and marked local erosion. Walvis Bay on the arid west coast of southern Africa is a good example of this type of climate, because whereas the average annual rainfall is 22·5 mm, the maximum recorded daily rainfall (1916–50) is 30·0 mm. At Karachi where the average annual rainfall is 193 mm, the maximum recorded daily rainfall (1875–1940) is 203 mm.

The Trade Winds

Trade winds occur on the equatorial side of the sub-tropical anticyclones and occupy the bulk of the tropics. Broadly speaking, the trade winds blow from ENE in the northern hemisphere and from ESE in the southern. They are marked by extreme constancy in both speed and direction for in no other climatic regime on earth do the winds blow so steadily, the only interruptions in the flow occurring when a rare atmospheric disturbance forms. There tends to be a certain monotony about the weather of the trade winds since the steadiness of the trades reflects the permanence of the sub-tropical anticyclones, which tend to be most intense in winter, making the trade winds strongest in winter and weakest in summer. All tropical oceans have extensive areas of trade winds except the northern Indian Ocean.

The trade-wind inversion, which has already been mentioned, is one of the most important climatological factors operating in the tropics. During the summer of 1856 an expedition under the direction of C. Piazzi-Smyth visited the island of Teneriffa in the Canary Islands. The purpose of their visit was to make astronomical measurements from the top of the Peak of Teneriffa. On two of the trips up and down the 3 000-metre mountain, Piazzi-Smyth carefully measured the temperature, moisture and wind, and found a temperature inversion with very dry air above it. He observed that the inversion was not located at the top of the north-easterly trade regime, but that it was situated in the very middle of the current and thus could not be explained as a boundary between

two streams of air from different directions. He also noticed that the top of the cloud layer corresponded with the base of the inversion, an observation which has been confirmed by aircraft and soundings with instrumented balloons (radiosondes).

An interesting feature of the trade inversion is the warmth and dryness of the air above the inversion which is caused by the broad-scale subsidence of the atmosphere in the sub-tropical anticyclones. It has already been stated that this subsiding air does not reach the surface but normally spreads out over a layer of relatively cool air at the surface. In the case of the sub-tropical highs the subsiding air normally meets a low-level stream of maritime air and the inversion forms at the meeting point of these two strata, both of which flow equatorward. The height of the inversion base is a measure of the depth to which the upper current has been able to penetrate downward towards the surface.

Subsidence is most marked at the eastern ends of

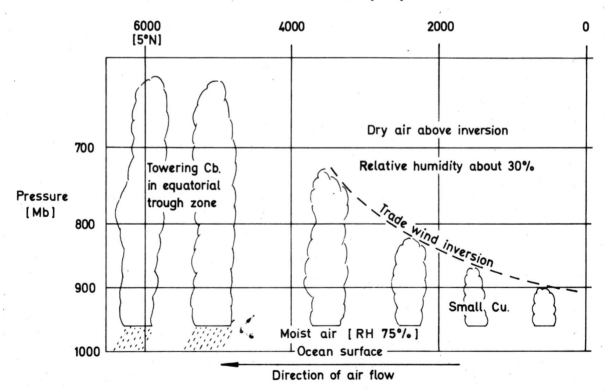

Figure 6.1 Schematic cross-section along trade winds flowing towards the equator. Northern hemisphere case. Compare with Figure 3.2.

the sub-tropical anticyclonic cells, that is to say along the desert cold-water coasts of the western edges of America and Africa, for it is here that the trade inversion is at its lowest. Normally as we follow (Figure 6.1) the trade winds towards the equator, the trade inversion increases in altitude and conditions become less arid. The typical cloud of the trade winds is the cumulus cloud, which is formed by warm bubbles or 'thermals' rising from the surface. The sky between the cumulus clouds is clear, and so large amounts of solar radiation reach the surface. Over the oceans the intense radiation evaporates water which is carried aloft by the thermals and eventually distributed throughout the layer below the trade inversion. The result is that the layer below the inversion becomes more moist as the trade wind nears the equator and the continual convection in the cool surface layer forces the trade inversion to rise in height.

The very low inversion can have unfortunate effects on cities located in the tropics. For example, in summer the low-level inversion spreads over Los Angeles and very effectively stops any vertical exchange of air from taking place. The gaseous waste combustion products from industry and domestic households, plus the exhaust fumes from 2 million motor-cars rise into this layer of the atmosphere which is already naturally hazy. The result is the formation of the notorious 'smog' of Los Angeles, whose name is derived from a combination of the words 'smoke' and 'fog'. Other tropical cities can suffer from similar smogs due to the trade-wind inversion.

The Equatorial Trough

Trade winds from the northern and southern hemispheres meet in the equatorial trough, which is a shallow trough of low pressure, generally situated on or near the equator. Over the oceans it lies in the belt of the doldrums and has a north and south movement which follows the sun with a time-lag of one or two months. Latitudinal changes normally amount to only a few degrees during the year, except over the Indian Ocean where they amount to about 30°. Though the winds are calm or light easterly in many equatorial regions, semi-permanent equatorial westerlies are found over the eastern Indian Ocean, Indonesia and Malaysia.

The trade-wind inversion is carried by the flow of the trade winds into the equatorial trough; its height increases towards the equator and in the equatorial trough it often vanishes. The result is that, while convection is restricted in the trade winds, towering cumulus clouds can develop in the equatorial trough and they give rise to heavy showers. The equatorial trough is therefore a relatively wet, cloudy zone compared with the trade winds, but it must not be thought that the equatorial trough is a region of continuous cloud and rain. Very large areas can be dominated by the trade-wind inversion and have fine dry weather. The rain areas tend to appear in a rather irregular manner and they are sometimes nearly stationary, tending to form and die in roughly the same region.

Cloud and rain in the equatorial trough are often associated with the inter-tropical convergence zone (ITCZ) which is a relatively narrow zone into which airstreams converge. Since it has characteristics which are distinct from the fronts of higher latitudes, the term ITCZ is generally preferred to the alternative 'inter-tropical front'. Over the sea it occurs at or very near the latitude of maximum surface temperature, which is usually some distance off the equator, while over land it forms at the boundary between converging airstreams. Inspection of individual photographs from earth satellites show that the ITCZ can only rarely be identified as a long, unbroken band of heavy cloudiness. Rather, it is usually made of a number of 'cloud clusters', separated by large expanses of clear sky. The cloud clusters are marked by heavy rainfall (> 2 cm/day) and normally move slowly westward. Thus, in a climatological sense, the ITCZ may be considered to be the locus of westward propagating cloud clusters. Both single and double versions of the ITCZ are known, sometimes one convergence zone is observed near the equator while at others there are two marking the northern and southern limits of the equatorial trough.

The distribution of mean annual rainfall within the equatorial trough is extremely variable. Rainfall maxima are observed over the equatorial regions of South America and Africa, and also over the equatorial islands from Indonesia to the Carolines. In contrast rainfall minima tend to exist over the oceans. Variations within the trough become clearer if the annual percentage of days with thunder are considered, since thunderstorm peaks are found over South America, Africa and Indonesia, while they are comparatively rare over the oceans. The bulk of the rainfall over the equatorial continents and Indonesia/Malaysia falls from thunderstorms, but this is not true of the oceans. Thus thunder is recorded at Buitenzorg on 322 days a year, Jakarta on 136, Surabaya on 73 and at Singapore on 41 days, but in contrast Fanning Island (Line Islands) with an annual average rainfall of 1 992 mm, has only one thunderstorm day a year.

Rain-forming Disturbances

In the equatorial trough rainfall is not particularly frequent, while in the trade winds proper, the sky will contain only small cumulus clouds and it may be rainless for many days. Only when some form of major synoptic disturbance exists does the sky become cloudy and rain form. Much of the rain in the tropical world falls from convective clouds, that is from large cumulus or cumulonimbus. In the most severe storms these clouds will be packed side by side to give continuous rain and cloud, but more usually there are wide gaps between the clouds which lead to a rather random and discontinuous rainfall distribution. Thus some places have heavy rainfall, while others nearby have little or no rain. This is in complete contrast to the continuous cloud and widespread rainfall so often observed in middle latitudes. Tropical synoptic systems can be organized in a variety of ways, but it is a common feature that they form within a uniform air mass, for the clash of air masses, so common in middle latitudes rarely gives rise to atmospheric disturbances within the tropics. Near the equator organized horizontal rotation is unusual because of the low value of the Coriolis parameter, but poleward of about 5° latitude, closed rotations can and do occur. This means that there are a whole series of synoptic disturbances that form in the sub-tropics but cannot affect the equatorial zone.

The simplest and most common form of ascent visible in the tropical sky is the cumulus cloud, which is the most common tropical cloud, and forms the outstanding element of the cloud scenery. Cumulus clouds are formed by convection, that is by bubbles of warm air or thermals rising through the atmosphere. To initiate convection, a source of energy is required, the best known of which is the differential heating of land in the day-time by the sun. Over most land areas the ground is at least slightly hilly, and vegetation and soil types vary, so the absorption of

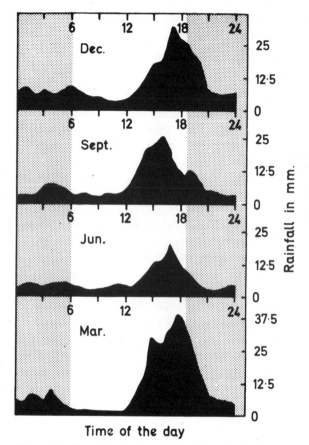

Figure 6.2 Diurnal variation of rainfall at Kuala Lumpur. Time refers to local time. (After Dale, 1960).

sun's rays is not uniform. Clouds will tend to form where the heating is strongest, where the general airstream is forced upward in hilly or mountainous country, or where convergence takes place in an airstream. Solar radiation imposes a diurnal rhythm on both cloud formation and rainfall, and this is particularly so in the equatorial trough zone, where cloud development tends to reach a maximum late in the afternoon. Figure 6.2 which illustrates the diurnal variation of rainfall at Kuala Lumpur, clearly shows a rainfall peak in the late afternoon—a common enough occurrence in the equatorial trough, but there are exceptions.

In the equatorial trough, rainfall is often associated with minor convergence zones, containing many cumulonimbus clouds and these are extremely difficult to examine, but in the sub-tropics there are two well-known and easily examined systems, the easterly wave and the tropical storm or cyclone.

Two types of wave disturbance in the tropical easterlies have been the subject of study; the easterly wave of the Caribbean and the equatorial wave of the Pacific. Historically the Caribbean wave was first studied, and the model (Figure 6.3) was then applied elsewhere. In the easterly flow over the Caribbean, shallow waves form and move slowly west at a speed of 5 to 8 m/s. They are about 15° latitude across, and the easterly winds flow through them from east to west. Typically in these waves, subsidence is found ahead of the trough line and ascent in the rear. The wave structure in the easterlies is weakest at sea-level and increases in intensity with height up to about 4 000 metres, but above this level it again becomes weaker. About 300 km ahead of the wave trough the trade-wind inversion reaches its lowest level and exceptionally fine weather prevails. The altitude of the trade-wind inversion and therefore the depth of the moist layer rise rapidly near the trough line, and the moist layer attains a maximum depth of well above 6 km in the zone of ascent behind the trough. Here are found large squall lines, intense rain and rows of cumulonimbus clouds. It should not be thought, however, that this region east of the trough line, even in the case of very strong waves, is completely covered by cumulonimbus clouds for wide zones of subsidence exist between individual clouds, and the cumulonimbus clouds are often organized into lines or rows.

Easterly waves are well known in the Caribbean and the west central Pacific. It would appear that many waves over the Caribbean have their origins

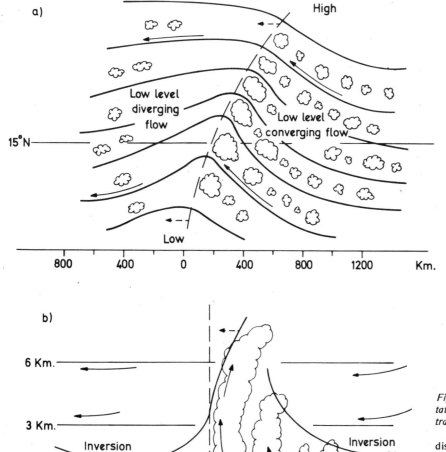

Figure 6.3 Schematic representation of an easterly wave in the trade winds.
Upper. General plan of the disturbance. The wind blows approximately along the isobars which are shown by thin continuous lines.
Lower. Cross-section.

over the African coast. The surges of the trade wind in the China Sea are similar to easterly waves, but there is no record of a persistent travelling wave in the China Sea.

Storms with closed circulation systems can occur anywhere within the tropics outside of the zone 5°N to 5°S. These systems can vary from slowly circulating masses of air with scattered cumulonimbus clouds to violent and severe storms. Although differences between the various tropical low pressure systems are not easily definable, the World Meteorological Organization classifies low pressure systems as follows.

(a) A tropical depression has low pressure within a few closed isobars, and either lacks marked circulation or has winds below 17 m/s.

(b) A tropical storm is one with several closed isobars and a wind circulation from 17 to 32 m/s.

(c) A tropical cyclone is a cyclone of tropical origin of small diameter (some hundreds of kilometres) with minimum surface pressure in some cases less than 900 mb, very violent winds, and torrential rain, sometimes accompanied by thunderstorms. It usually contains a central region, known as the 'eye' of the storm, with a diameter of the order of some tens of kilometres, and with light winds and more or less lightly clouded sky.

Tropical cyclones are given a variety of regional names. In the south-west Pacific and Bay of Bengal the name 'tropical cyclone' is in use; they are known as 'typhoons' in the China Sea, as 'willy-willies' over western Australia, and as 'hurricanes' in the West Indies and in the South Indian Ocean.

They require a plentiful supply of moisture, and for this reason their incidence is limited to the regions where the highest sea-surface temperatures are found, that is to the western regions of the tropical oceans in the late summer. Not all incipient tropical low pressure systems become tropical cyclones, for many remain as weak closed circulations with moderate rain and light winds. Though weak tropical disturbances are relatively common, tropical cyclones are rare, since even in an active year, only seldom will the total number of tropical cyclones in the northern hemisphere exceed fifty, whereas in contrast, about twenty depressions occur almost every day outside of the tropics in winter.

Tropical cyclones (Figure 6.4) are regions of intense ascent which usually takes place in cumulonimbus clouds arranged in spirals converging on the central eye. The spirals, sometimes hundreds of kilometres long, are at most a few kilometres wide, and the distance between spirals is 50 to 80 km, decreasing toward the central eye. Therefore only a small fraction of the tropical cyclone, not more than 10 per

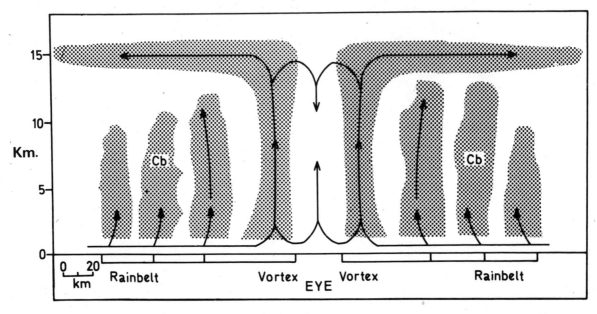

Figure 6.4 Schematic section through a tropical cyclone. The cumulonimbus clouds are arranged in spiralling bands. Arrows indicate horizontal and vertical wind components. (After Flohn, 1969.)

cent, often contains the rising air currents giving rise to most of the condensation and rainfall. Close to the centre, the clouds form a ring around the eye, and here the strongest wind and heaviest rainfall combine to produce the storm's full fury. The central eye is a most interesting feature, because in the interior of the eye which can be up to 20 km across, the winds decrease quickly and the heavy rains cease. Modern research has shown that the eye is actually a region of subsiding motion at the centre of the storm and it appears to be essential to the formation of a tropical cyclone, since tropical depressions and tropical storms do not normally contain a central eye with subsiding air. One of the first signs that a tropical storm is going to turn into a tropical cyclone is the formation of a central eye.

The energy to drive the wind system of the tropical cyclone comes mainly from the latent heat released by condensing water vapour which is normally evaporated from the ocean close to the storm. Tropical cyclones are therefore normally restricted to the warmest parts of the tropical oceans, for rainfall rates near the centre of a tropical cyclone can reach 500 mm per day, and so a continuous supply of water-vapour is required to keep the storm operating.

Despite the strong winds, the rate of movement of a tropical cyclone is only about 5 to 10 m/s. In the northern hemisphere the storm will drift towards the north-west at first but when the centre passes latitude 20°N, the speed of travel increases and the direction usually changes to a more north-easterly path. The point·at which the direction of motion of the cyclone changes from being westerly to easterly is known as the 'point of recurvature'. The system then usually begins to fill up, that is to say, pressure rises in the centre. The recurvature of cyclone tracks in the southern hemisphere is, in the opposite sense, storms moving first towards the south-west and later towards the south-east. Cyclones, however, never cross the equator. If a tropical cyclone drifts across a tropical land mass, it decays very quickly, normally within one or two days and this is due to the cutting off of the supply of water-vapour.

Tropical cyclones normally form in oceanic areas where the sea-surface temperature is at least 26·5°C, such values only being found in regions of warm currents in late summer. So they are most frequent in the months from July to November in the northern hemisphere. The main development areas are indicated in Table 6.3.

The Monsoon Climates of Southern Asia

Aridity is the climatic norm of much of the subtropics, as in North Africa and the Middle East, but extensive summer rains make southern Asia an exception. These summer rains are associated with a general reversal of the north-east trade winds, the area being covered instead by south-west winds. The term monsoon, derived from the Arabic word 'mausim' meaning 'season', is often used to describe these markedly seasonal winds. Thus it is possible to talk about the south-west and north-east monsoon seasons, using wind direction as the main characteristic of the season. The characteristics of the monsoon climate are best developed in the Indian subcontinent but variations of the climate are found in the sub-tropical zone extending from Pakistan to the Philippine Islands. In the Indian subcontinent, the seasonal changes over large areas may be conveniently divided as follows:

(a) The season of the north-east monsoon;

    (i) January and February, winter season;

    (ii) March to May, hot weather season;

(b) The season of the south-west monsoon;

    (i) June to September, season of general rains;

    (ii) October to December, post-monsoon season.

January is a fine sunny month over most of India, since the generally north-easterly winds give little or no rain. Indeed, during the north-east monsoon season the area is dominated by high pressure and experiences a normal trade-wind type climate. The winter season merges gradually into the hot weather sea-

TABLE 6.3
TROPICAL CYCLONE DEVELOPMENT AREAS

Area	Percentage of World-wide total
N.E. Pacific	16
N.W. Pacific	36
Bay of Bengal	10
Arabian Sea	3
South Indian Ocean	10
Off N.W. Australian coast	3
South Pacific	11
N.W. Atlantic (including W. Caribbean and Gulf of Mexico.)	11

son, for in April and May the sun is nearly overhead, and the days become very hot, giving the maximum annual temperatures over much of the area. In June the south-west monsoon sets in, or, to use the more common term 'bursts'. This is the season of general cloud and rain, and because of the extensive cloud cover, the temperatures are lower than those of April and May. The monsoon rains are not continuous and lengthy breaks can occur. In October and November the south-west monsoon retreats, the sky clears, the sun shines again, and the temperature rises for a few weeks before falling to the winter minimum.

When upper winds are taken into account, it is found that the Asian monsoon is a fairly complex system. During the northern winter season, the subtropical westerly jet stream lies over southern Asia, with its core located at about 12 km altitude. It divides in the region of the Tibetan Plateau, with the weaker branch flowing to the north of the plateau, and the stronger to the south. The two branches merge to the east of the plateau and form an immense upper convergence zone over China causing general subsidence and fine weather. In May and June the sub-tropical jet stream over northern India slowly weakens and disintegrates, causing the main westerly flow to move north into central Asia. While this is occurring, an easterly jet stream mainly at about 14 km above sea-level builds up over the equatorial Indian Ocean and expands westward into Africa. The formation of the equatorial easterly jet stream is connected with the formation of an upper-level high pressure system over Tibet. In October the reverse process occurs; the equatorial easterly jet stream and the Tibetan high disintegrate while the sub-tropical westerly jet stream reforms over northern India. Thus there are reversals of wind direction in both the upper and lower troposphere, and in the northern summer the surface westerlies over India only extend up to about 4 km being overlain by powerful easterly winds.

The south-west monsoon season comprising the four months of June, July, August and September accounts for more than 75 per cent of the annual rainfall for most parts of India, except for peninsula India south of 15°N, which experiences its rainy season during the north-east monsoon months of October and November. The northern and north-eastern parts of India also experience precipitation associated with pre-monsoon thunderstorms which progressively increase from March to May, and similar travelling disturbances in the extra tropical westerlies give rise to light precipitation from December to May in the north-western parts of India.

Monsoon rainfall over India presents a variety of seasonal patterns, which are partly due to orography and partly to atmospheric factors. Stations (Figure 6.5) on the west coast show a steep increase in May and June followed by a gradual decline, in marked contrast to the gradual rise and fall revealed by the north Indian stations. Thus Mangalore has, on average, a sharp increase in rainfall commencing about 20 May and reaching a peak value by about the middle of June. Indeed, the steep rise in rainfall in May and June at stations on the west coast of India explains the origin of the expression 'burst of the monsoon on the Malabar coast', for by taking rainfall as the criterion, the date of onset of the south-west monsoon can be fixed with greater precision here than for the rest of the subcontinent. Stations on the west coast south of latitude 12°N, such as Trivandrum, reveal two rainfall peaks, one in June and the other in October, with the October maximum being lower than the one in June. On the east coast of India, stations between latitudes 15° and 19°N have their maximum rainfall in the month of October, those south of 15°N in November, and those north of 20°N in July. Stations in the interior of the peninsula show a number of interesting features; for example, Sholapur receives its maximum rainfall in September while Hyderabad has two rainfall peaks in July and September. In north India the rainfall decreases from east to west, the reverse of the gradient found in the south, and the south-west monsoon current does not reach Srinagar in the Kashmir Valley, since this station experiences more rainfall in the winter months in association with westerly disturbances than in the southwest monsoon months. Equally complex annual rainfall patterns are found in other parts of southern Asia, for example, in the southern part of the Malay Peninsula, rainfall is evenly distributed throughout the year with a tendency for a maximum between October and March. Proceeding north along the peninsula into Burma and Thailand it is found that a distinct dry season develops in the northern winter, most of the rain coming in the summer months. Similar comments can be made about much of southern China where there is a distinctive rainfall maximum over wide areas in the summer months.

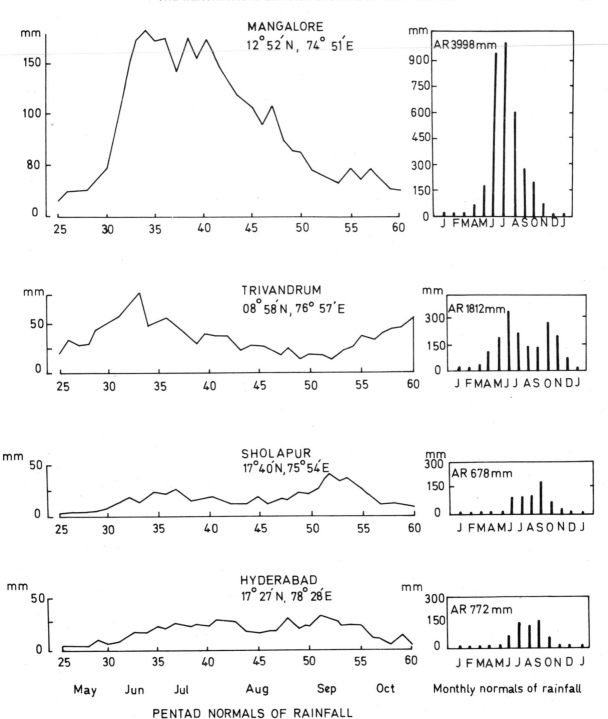

Figure 6.5 *Seasonal rainfall patterns at four stations in the Indian sub-continent.*

Left. Five-day mean rainfall. Pentad numbers are from the beginning of the year.

Right. Monthly mean rainfall and annual average rainfall. (After R. Ananthakrishnan, and P.J. Rajagopalachari (1964). 'Pattern of monsoon rainfall distribution over India and neigbourhood', *Proc. W.M.O., I.U.G.G. Symp. Trop. Meteorol* (Rotorua, New Zealand).)

Though rainfall is often termed monsoonal over much of southern Asia, it may not be directly connected with the Indian south-west monsoon. Rainfall is the result of vertical motions in humid air, and therefore to explain its seasonal variations it is necessary to consider seasonal changes in atmospheric humidity and ascent patterns. Ascending air cools because it is expanding under the influence of falling atmospheric pressure, and eventually the water-vapour it contains is condensed and forms cloud and rain.

During the north-east monsoon season the conditions over southern Asia are broadly similar to those found elsewhere in the tropics. A sub-tropical westerly jet stream is found over northern India, and associated with this, there is general subsidence causing generally dry weather. Over the equatorial Indian Ocean equatorial westerlies flow eastward between the Hadley cell circulations of the two hemispheres and meet equatorial easterlies from the Pacific over Indonesia. This creates a large convergence zone over Indonesia causing ascent and therefore a rainfall maximum at this time of the year.

Heating of the highlands and deserts of southern Asia by the sun in the spring causes the sub-tropical westerly jet stream over north India to disintegrate and be replaced by surface low pressure overlain by high pressure. This results in surface westerly winds over southern Asia, and these are overlain by high-level easterlies. Once the general subsidence is removed, tropical storms of various types can develop over the oceans and these are the cause of much of the monsoonal cloud and rainfall. From May to November, tropical cyclones form in the Bay of Bengal and move west or north-westwards into India causing widespread rainfall. The condensation of water-vapour in these storms releases latent heat which further intensifies the monsoonal circulation. It is found for instance, that the high-level easterly jet stream is most vigorous when there is heavy rainfall over India. Similarly, disturbances form in the Arabian Sea and bring rainfall to the west coast of India. During this season tropical cyclones (typhoons) develop over the Pacific and move north-westwards across the Philippines into southern China, and these are responsible, together with weaker disturbances, for much of the summer rainfall of these areas. While conditions north of the equator are highly disturbed, conditions on the equator and to the south are quiet and this is the season of minimum rainfall over much of Indonesia.

With reduced solar heating over southern Asia in the autumn, the atmosphere returns to normal, the sub-tropical westerly jet stream becomes re-established and its associated subsidence rapidly damps down any tendency for disturbances to form.

The Climate of Equatorial South-East Asia

Since the duration of daylight varies only slightly from 12 hours throughout the year near the equator, annual changes in radiation are usually small. The high rainfall supplies abundant moisture for evaporation, which uses up large amounts of net radiation, therefore the temperatures tend to be not only nearly uniform throughout the year but also rather low when compared with the arid tropics. The thermal regime of the region is characterized by general uniformity, the sea-level temperatures being almost constant throughout the whole area and throughout the year. At any given site the largest variations in temperature are between day and night, rather than between seasons. Mean annual sea-level temperatures are everywhere close to 26°C.

Diurnal temperature variations follow the same pattern in most parts of Malaysia and Indonesia; the daily minimum is reached between 5 and 7 a.m. local time, and the maximum between 12 a.m. and 3 p.m. A fairly rapid rise in temperature takes place during the first 6 or 7 hours of daylight, and after reaching the maximum the temperature falls again, fairly rapidly in the late afternoon and slowly through the night. The annual temperature range at Singapore is about 2·2°C while the diurnal range reaches 6·2°C, but annual ranges increase with distance from the equator.

Temperature falls with altitude at a rate of between 0·6°C and 0·5°C per 100 m, so permanent snow is only found on the most lofty peaks, which are in New Guinea. Nevertheless, as is explained in Chapter 4, frost can occur at relatively low altitudes in mountainous regions.

While temperatures largely reflect the radiation climate, precipitation is the result of atmospheric motions, which vary widely in both space and time. Variations in atmospheric flow patterns lead to large variations in precipitation along the equator. Precipitation maxima are observed over the equatorial regions of South America and Africa, and also over the equatorial islands and peninsulas from Indonesia to the Carolines, while minima tend to exist over the oceans. Indeed, the precipitation over parts of the equatorial eastern Pacific is similar in amount to that observed in the sub-tropical deserts.

Surface air over South-East Asia is always warm and humid, and therefore the annual variation of precipitation is governed by the movement of convergence zones, which in turn are controlled by events outside of the region. Thus a rainfall maximum exists over Indonesia during the northern winter because of the massive convergence between the northern and southern trades, but in the northern summer there is slight divergence and consequently a rainfall minimum.

Annual precipitation changes over Malaysia and Indonesia largely reflect changes in the sub-tropics and in particular the generation of the south-west monsoon over India. During the northern winter large sub-tropical anticyclones are often found over Australia and the Thailand–Indo-China region, and the trades flowing out of them converge over South-East Asia. In the northern summer the sub-tropical highs over Australia remain, but those over southern Asia vanish and are replaced by extensive south-westerly winds and moving low pressure systems.

Over the equatorial continents, the bulk of the rainfall comes from showers or thunderstorms, and in this respect the scattered peninsulas and islands of

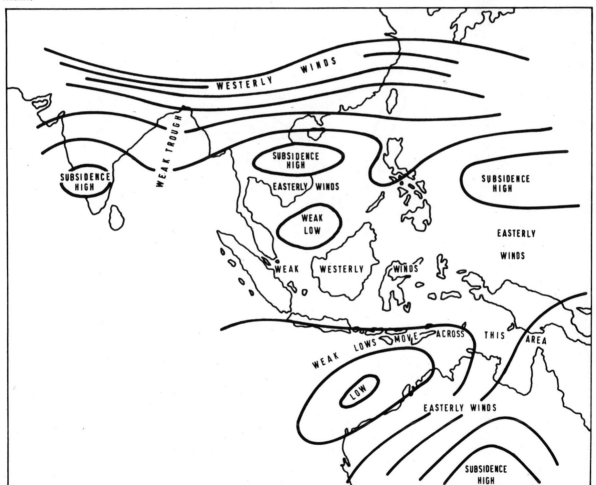

Figure 6.6 Graphical illustration of the structure of the atmosphere over South-east Asia during January. The solid lines are typical contours of the 700 mb (about 3 km) surface and are drawn at 15 m intervals. They illustrate the positions of the major areas of high and low pressure rather

better than surface charts. Much of Indonesia is dominated by low pressure systems which cause extensive rainfall. Similarly, the disturbances in the South China Sea bring rainfall to the eastern part of the Malay Peninsula. Much of southern Asia is dominated by anticyclones and has dry weather.

South-East Asia act as a continent. In general, annual
rainfall totals exceed 2 000 mm, and stations near the
equator tend to have it almost evenly distributed
throughout the year, but further away from the equator the rainfall becomes more seasonal with one pronounced maximum. The months from November to
February (Figure 6.6) form the wet season over much
of South-East Asia, when active convergence is taking
place and satellite pictures show extensive cloud
cover. The main exception to this is in the north of
the region, where the maximum rainfall occurs during
the south-west monsoon season. During these months
weak stationary lows frequently form in the South
China Sea and cyclonic activity often occurs over
northern Australia and southern Indonesia. The mean

inter-tropical convergence zone occurs over the Java
Sea during January, and from there it runs south-east
to the Arafura Sea and the Torres Strait.

March, April and May represent the transition period from the wet to the dry season over much of
Indonesia and Malaysia. As the season advances the
sub-tropical anticyclone over Australia drifts north
bringing dry easterly winds to within about 5°
of the equator.

June, July and August (Figure 6.7) form the dry
season over most of the region, but there are some
very marked meteorological contrasts across the equator. The sub-tropical anticyclone over Australia is
now near to its most northerly position, and over
much of Indonesia easterly winds exist which are nor-

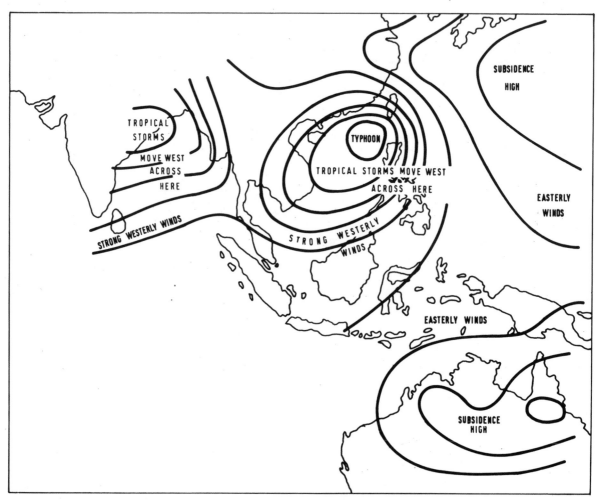

Figure 6.7 Graphical illustration of the structure of the atmosphere over South-east Asia during July. Key as for
Figure 6.6. While much of southern Asia is dominated by

disturbances bringing rain, Indonesia is under the influence of
the Australian anticyclone and has generally dry weather.

mally very dry, indicating subsidence, in contrast to the moist south-westerlies to the north of the equator. This is the season of the south-west monsoon over most of southern Asia, and deep tropical storms move westward across the Philippines towards China. Indeed, this is the period of maximum rainfall over Thailand and parts of northern Peninsular Malaysia.

The transition back to the wet season occurs in September, October and November. This period marks the decline of the south-west monsoon over southern Asia and the building of anticyclones across China and later across India. During October the Australian high is still near to its extreme northern position, but it moves south in November as cyclonic activity increases over northern Australia. Intense lows develop over the South China Sea, particularly in November, and the north-easterly winds associated with them bring heavy rain to the east coast of Peninsular Malaysia.

Rainfall distribution patterns tend to reflect the major features of the topography. In Peninsular Malaysia there are two particularly wet belts (annual rainfall above 3 000 mm): one running the full length of the South China Sea side of the eastern mountains, and the other confined to the northern part of the peninsula, on the Malacca Straits' side of the western mountain ranges. The axes of the high rainfall belts do not, in general, coincide with the highest land, but rather with the foothills of the mountains. For example, the axes of the western wet belt in Peninsular Malaysia appear to approximate with an altitude of between 150 m and 350 m, beyond which the rainfall decreases with increasing height. The eastern wet belt covers the mountain slopes and foothills facing the South China Sea and extends across the lowlands to the coast. Indeed, one of the wettest places in the eastern part of the peninsula is situated on the lowlands at less than 130 m above sea-level and only 19 km from the coast. While the coastal lowlands have large annual rainfalls, lowland areas in interior valleys can form marked dry areas (annual rainfall below 1 700 mm). Rainfall distributions are observed in the Indonesian Islands which are similar to those found in the Malay Peninsula.

The importance of atmospheric motions in generating rainfall is well illustrated by the diurnal variation of rainfall in Peninsular Malaysia. The diurnal march of temperature is similar over the whole peninsula, but a variety of diurnal rainfall regimes are observed. The common one (Figure 6.2) is for a maximum of rainfall between midday and midnight, the rest of the day being relatively dry. In August, the heaviest rain falls between midnight and dawn along the coast near Malacca, and this is caused by the interaction of land and sea breezes from Sumatra and Peninsular Malaysia over the Malacca Strait. At night the convergence of the land breezes causes upward motion over the Malacca Strait; and since the Sumatra land breeze reinforces and the Malaysian land breeze opposes the westerly synoptic wind, the convergence zone in which showers develop probably lies close to the Malaysian coast. Showers drift over the coast and cause the night rainfall maximum, nevertheless the showers decrease during the day when the sea breezes set in, generating divergence and sinking in the Strait and along the coast. Similarly, night rainfall maxima exist in some inland valleys, and these are caused by katabatic winds producing low-level convergence. During the day, anabatic winds cause low-level divergence in the valleys and a rainfall minimum.

FURTHER READING FOR CHAPTERS 5 AND 6

Chang Jen-Hu (1972). *Atmosphere Circulation Systems and Climates* (Oriental Publishing Co., Hawaii).

Crowe, P.R. (1971). *Concepts in Climatology* (Longman, London).

Dale, W.L. (1959). 'The rainfall of Malaya, Part 1'. *Journal of Tropical Geography*, 13, p. 23.

———(1960). 'The rainfall of Malaya, Part 2'. *Journal of Tropical Geography*, 14, p. 11.

———(1963). 'Surface temperatures in Malaya'. *Journal of Tropical Geography*, 17, p. 57.

———(1964). 'Sunshine in Malaya'. *Journal of Tropical Geography*, 19, p. 20.

Flohn, H. (1969). *Climate and Weather* (Weidenfeld and Nicolson, London).

Lockwood, J.G. (1974). *World Climatology: An Environmental Approach* (Edward Arnold, London).

Ooi Jin Bee and Chia Lin Sien (1974). *The Climate of West Malaysia and Singapore* (Oxford University Press, Singapore).

Riehl, H. (1965). *Introduction to the Atmosphere* (McGraw-Hill, New York).

Sutcliffe, R.C. (1966). *Weather and Climate* (Weidenfeld and Nicolson, London).

Weyl, P.K. (1970). *Oceanography, an Introduction to the Marine Environment* (Wiley, New York).

PART D

Water in the Tropical Environment

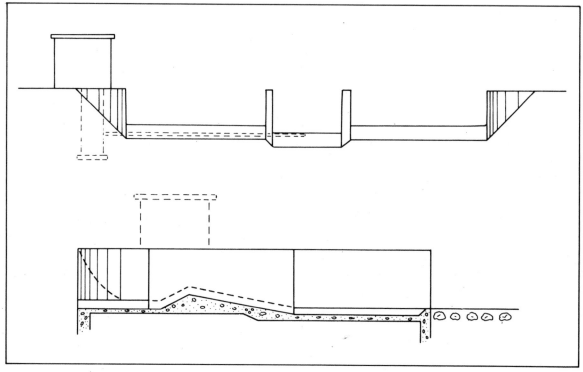

Plan of a small weir.

7 The Hydrological Cycle

HYDROLOGY is defined as the study of the incidence and properties of water on and within the ground, including that held in rivers and lakes. It comprises studies of rainfall, evaporation, run-off, groundwater, soil moisture, the hydrological balance, snow and ice accumulation and the chemistry of natural waters. The related science of glaciology is defined as the study of the distribution and behaviour of snow and

Figure 7.1 The hydrological cycle. The hydrological cycle can be considered as a very complex cascading system, with many moisture stores and moisture transfers, but all the stages of the hydrological cycle may not operate at the same time.

Residence times of water in various stores are listed below, where the residence time is the time it would take to double the normal content, given normal rates of inflow and no losses.

Oceans	about 44,000 years
Atmosphere	about 10 days
Vegetation canopies	few minutes to few hours
Soil	several weeks to several months
Ground water	several months to many years

Rates of transfer from one store to another are also listed below:

Moisture advection by winds from oceans to continental interiors	few days
River flow from continental interiors to oceans	few weeks to several months
Evapotranspiration from moist soil and plants to atmosphere	few hours
Ground water flow from continents to oceans	very slow, many years
Precipitation from atmosphere to surface	few hours

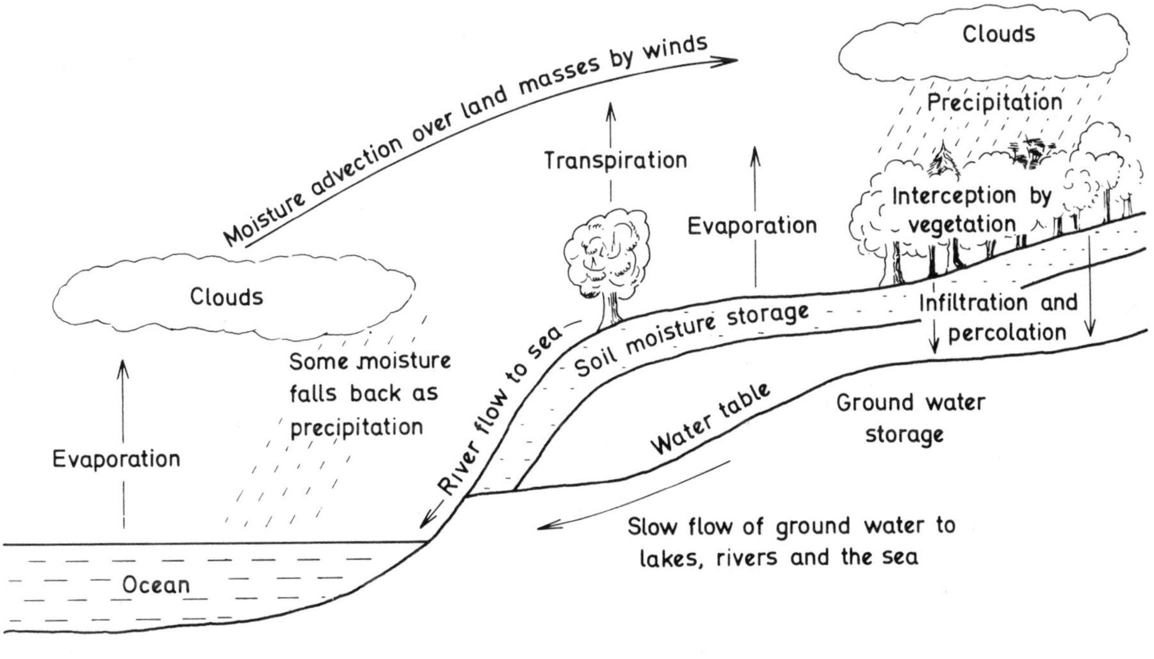

ice on the earth's surface, and it is therefore considered in this section together with hydrology. The full cycle of events through which water passes in the earth-atmosphere system is best illustrated in terms of the hydrological cycle (Figure 7.1), which describes the circulation of water from the oceans, through the atmosphere back to the oceans, or to the land and thence to the oceans again by overland and subterranean routes. Water in the oceans evaporates under the influence of solar radiation and the resulting clouds of water-vapour are transported by the atmospheric circulation to the land areas, where precipitation may occur in the form of rain, hail or snow. Some of this precipitation will infiltrate into the soil and then percolate into the saturated zone beneath the water-table, from where it flows slowly through aquifers to river channels or sometimes directly to the sea. The water remaining on the surface will partly evaporate back to the atmosphere and partly form surface run-off into rivers, which eventually run into the oceans. To the moisture which has evaporated directly back to the atmosphere from the soil surface, must be added a further loss of water in the soil via transpiration through plants. Not all the above stages will necessarily occur in any particular example of the hydrological cycle or at any particular place or time. During droughts the cycle may appear to have stopped, whereas during floods it may seem to be

continuous, and yet both of these phenomena can occur in the same place at different seasons.

The nature of the hydrological cycle can be further explored with the aid of Figure 7.2, which represents the atmospheric water-balance of a sub-continental area. Suppose that through the windward vertical boundary of the sub-continental region an amount of moisture arrives which is equivalent to an amount A per unit area and time. This process of transfer of water-vapour (or of any air mass property) by virtue of atmospheric motion is termed advection. Now, suppose that within the bounded area an amount of precipitation τ falls during the same time, which is made up of water of an amount τ_A formed of advected moisture, plus an amount τ_E due to evaporation within the bounded area. Assuming, therefore, that the precipitation comes from water-vapour of mixed origin, that is it is formed from local and advected vapour, it is possible to write:

$$\tau = \tau_A + \tau_E$$

Part of the moisture evaporating from the ground (E_t) goes to form the local precipitation τ_E, while an amount C is carried beyond the boundaries of the region. The quantity C is often called the atmospheric run-off:

$$E_t = \tau_E + C$$

Figure 7.2 Continental water-balance

A	moisture advection per unit area and time.
τ	total precipitation
τ_A	component of precipitation formed of advective moisture
r_E	component of precipitation formed of water evaporated over continent
E_t	evapotranspiration
C	atmospheric run-off
f	river flow

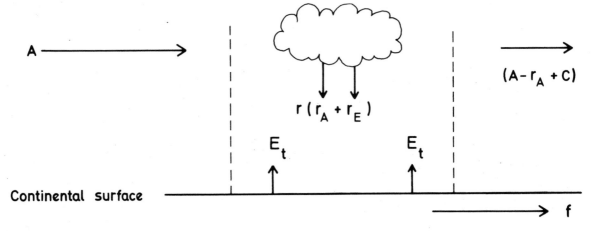

In this way, moisture is carried beyond the boundaries of the region either by advection plus atmospheric run-off $(A - \tau_A + C)$, or by stream flow f. The long-term average moisture balance for the region yields the following relationships:

$$\tau = E_t + f$$
$$\tau_A = C + f$$

The preceding discussion represents an attempt to start to describe the hydrological cycle in a quantitative manner, but without actual values for the various components it is difficult to assess their relative importance. For instance, if local evaporation is of greater significance than advected moisture, then neighbouring conditions may be of great importance in controlling rainfall, but in contrast, if most rainfall is formed of oceanic water, then local conditions will be of less interest. Relief always greatly influences the distribution of rainfall, but if it is mostly formed of locally evaporated water then it should be possible to alter the amount of rainfall by changing the rate of evapotranspiration. This could be done by cutting down trees in wooded regions or by planting trees and creating large lakes in semi-desert areas. Since man does modify the surface of the earth significantly, it is of great importance to study the hydrological cycle in a quantitative manner.

At any given time the earth's atmosphere contains on average an amount of water-vapour, which if it were all condensed and deposited on the surface of the earth, would stand to a depth of about 25 mm. In the actual atmosphere, water-vapour is far from evenly distributed over the earth's surface, the greatest amounts being found in the equatorial zone. Soil water storage is of rather more importance, since the soil can hold up to 100–150 mm of water, while groundwater storage can amount to the equivalent of several metres of rainfall. Water storage in the oceans is vast and in this context their storage capacity can be considered to be infinite. The amount of water held in the atmosphere, at any given time, in the form of water-vapour is small compared with the storage capacities of the soil, deep rocks and oceans, and so the main hydrological function of the atmosphere is to transport water-vapour; in fact, over a number of years there is an exact global balance between evaporation and precipitation.

It is difficult to distinguish between the two components of the water which forms the precipitation over the continents, but this has been managed by Begemann and Libby (1957), who by using water isotopes determined the ratio of maritime water to land water in the upper Mississippi Valley. Because of the tests of hydrogen-bomb weapons in the early 1950s, the quantitative ratio between the different water isotopes over the terrestrial globe has varied, and was different at times after the weapon tests for maritime water and land water. Water is a compound of oxygen and hydrogen and it is an isotope of hydrogen called tritium which is of particular interest in the water. Tritium is continuously formed in the high atmosphere by the action of cosmic radiation, and has also been injected into the atmosphere by thermonuclear explosions. By checking the isotopic composition of precipitation, it is possible to determine what percentage is derived from water-vapour transported from the ocean and what percentage came from vapour formed on the continent. Begemann and Libby studied the northern Mississippi Valley, U.S.A. where they concluded that one-third of the average precipitation is formed of re-evaporated water and two-thirds of ocean water.

Since the average annual precipitation in the northern Mississippi Valley is 770 mm, it follows that about 520 mm is ocean water and about 250 mm is re-evaporated rain-water. Annual run-off in rivers and streams is equivalent to about 280 mm of rainfall, and since annual precipitation must be balanced by annual evapotranspiration and run-off, this leaves 490 mm as the annual evapotranspiration. In order that the local evapotranspiration should supply only one-third of the atmospheric water-vapour, the winds must carry in from the oceans the equivalent of about 1 000 mm of rainfall per year, of which only 520 mm is actually precipitated as rain in the valley. The remaining 480 mm together with 240 mm of locally evaporated water is carried by the winds back from the continent to the oceans.

These values for the Mississippi Valley are interesting because they suggest that much of the rainfall over the continents consists of water which has been directly evaporated from the oceans, and this conclusion appears to hold even in the arid regions of Central Asia, where the proportion of precipitation derived from locally evaporated water-vapour is only about 4 per cent. It is therefore unlikely that local changes in the nature of the land surface, such as the removal of forest, will have any significant influence on local rainfall, but instead variations in rainfall can be traced back to variations in the atmospheric circulation and to sea-surface temperatures in the main oceanic source regions for water-vapour. Research indicates that the main sources of water-vapour are the oceans in the sub-tropics, where there are continuously clear skies and large amounts of solar radiation available to evaporate water. Thus South-East Asia receives much of its rainfall in the form of water which has evaporated in the sub-tropical Indian and Pacific Oceans, while the rainfall in Europe consists largely of water from the sub-tropical Atlantic.

On average the water-vapour in the air at any one time represents about one-fortieth part of the annual rainfall, or about ten days' supply of water. As rainfall consists of moisture which has been advected into the area by the winds, it is therefore reasonable to expect a marked correlation to exist between wind systems and rainfall. It is observed that the average water-vapour content of the atmosphere does not correlate very clearly with annual rainfall, and this is because rainfall is caused by weather systems and storms in the atmosphere. Consequently annual rainfall distributions reflect both the incidence of weather systems and also the atmospheric moisture content. Rainfall is high over South-East Asia because the winds carry moisture into the area and the incidence of weather systems in the form of thunderstorms is high. The deserts of North Africa receive little rainfall because weather systems are infrequent and so rainclouds cannot form. Similar rainless areas are found over the oceans, the central eastern Pacific being a good example, and this clearly illustrates the fact that precipitation distributions are controlled by atmospheric flow patterns.

Precipitation

Precipitation which is the source of all fresh water on the earth's surface, may take the form of rain, snow, hail, frost or dew. Over most of the earth's surface the most important form of precipitation is rainfall, though dew may be important locally in arid regions and snowfall on high mountains and in polar regions. Individual storms contribute widely differing amounts of precipitation to the earth's surface, but the percentage of annual precipitation supplied by large and small storms appears to be fairly constant irrespective of location on the globe, for it is found that approximately half of the annual precipitation over any given area is contributed by about one-quarter of the storms experienced. In contrast, about half the storms experienced contribute only about one-quarter of the annual precipitation total. Most of the annual precipitation total is caused, therefore, by a comparatively small number of major storms, while the great majority of storms produce rainfalls which are not hydrologically significant.

Four mechanisms are necessary for the production of significant precipitation:

(a) a lifting mechanism to produce cooling of the air;

(b) a mechanism to produce condensation of water-vapour and formation of cloud droplets;

(c) a mechanism to produce growth of cloud droplets to sizes capable of falling to the ground against the rising air currents implied by condition (a);

(d) a mechanism to produce sufficient accumulation of water-vapour in the storm to account for the observed precipitation rates.

These four mechanisms will now be considered in detail. The maximum amount of water-vapour that the air can hold at any given temperature (Figure 7.3) is fixed, but the moisture capacity of the air does increase as temperature increases. The relative humidity of the air may be considered as the percentage saturation, so when the humidity is 100 per cent the air is completely saturated with water-vapour at that temperature. If the absolute water-vapour content of the air remains constant but the temperature increases, then the relative humidity will fall because the moisture-holding capacity of the air has increased. Similarly, if a sample of air is cooled its relative humidity will increase until it becomes saturated, and if further cooling then takes place the air will become supersaturated with water-vapour and some will condense as water droplets. The only known mechanism capable of producing the degree and rate of cooling needed to account for heavy rainfall is the vertical ascent of air accompanied by adiabatic cooling. Thus the variation of precipitation in space and time is largely determined by the spatial and temporal variations in the vertical motion of air. The vertical motions observed in the atmosphere are largely a result of dynamic process within the atmosphere itself and the interactions of the atmosphere with the underlying surface. They are associated with weather

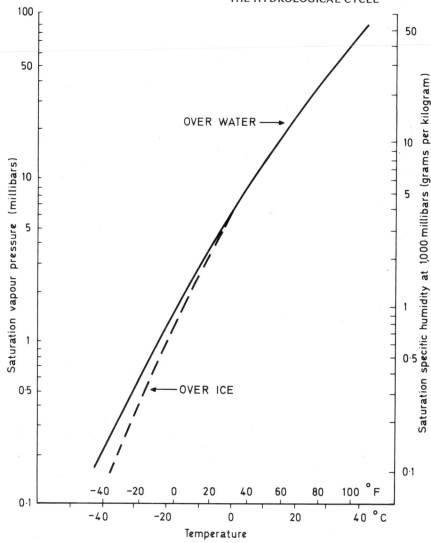

Figure 7.3 Saturation vapour pressure at various temperatures. To a high degree of approximation the capacity of the air to hold water in the form of water vapour depends only on temperature. A sample of moist air is said to be saturated if it can exist in contact with a plane surface of pure water or ice at the temperature of the air without evaporation or condensation. Below 0°C there are two curves, one for supercooled water, the other for ice.

systems of various scales ranging from thunderstorms to tropical cyclones and temperate latitude depressions. Vertical air motions in the atmosphere which give rise to significant rainfall may be classified as convective, banded and general.

Convective vertical motion occurs in cells with a diameter of about 1·5 to 8 km and a vertical extent from about 1 000 to more than 15 000 m. These cells are visible in the atmosphere as cumulus and cumulonimbus clouds and give rise to the shower type of precipitation. It is a characteristic of convective activity that the resulting precipitation is local in extent and of only short duration. Over relatively featureless terrain the spatial development and movement of convective cells is essentially random, but when the surface has marked orographic relief there is a tendency for convective cells to develop over preferred areas. This can be observed in South-East Asia where the many mountainous islands result in a complex distribution of convective cells which nevertheless are related to the orography.

Banded vertical motion, which occurs over a zone from a few kilometres to about 80 km in width and several hundred kilometres in length, is usually associated with fronts and convergence zones, or is the result of ascent over a mountain by an airstream. These systems may cause precipitation lasting for just a few hours or for several days.

General vertical motion is associated with large-scale weather systems such as tropical cyclones.

Broad-scale ascent takes place in these systems and the resulting precipitation can extend over the area of a medium-sized country. Organized convection bands commonly occur within tropical cyclones but the isohyetal patterns for the complete duration of such storms frequently lack significant cellular features. Storms of this type can have life cycles of many days and travel many thousands of kilometres.

As a broad generalization it can be stated that the smaller the horizontal extent of the vertical motion, the greater the velocity of the motion. Thus very high vertical motions are found in cumulonimbus clouds, while vertical motion over large areas is usually rather slow. Intense precipitation tends therefore to be associated with convective activity, since the vertical motions are rapid, but the intensity of the precipitation tends to decrease as the horizontal scale of the system increases. It is a characteristic of shower precipitation that it is of short duration and that it only covers a few square kilometres but can reach intensities of 10 to 15 cm per hour. Such intensities are unusual in larger-scale systems, which give gentler falls over larger areas for longer durations. Nevertheless, the total precipitation during large-scale storms can be greater than that in a thunderstorm because of their longer duration.

The cooling associated with rising air parcels produces a relative humidity of 100 per cent, and further cooling will produce supersaturation. It has been found in the laboratory that pure, dust-free air can have a relative humidity of considerably above 100 per cent without condensation, while it is observed that condensation takes place in the atmosphere with very small supersaturations, implying that atmospheric air differs in some way from pure, dust-free laboratory air. Investigations have shown that condensation in the atmosphere takes place on small dust particles having an affinity for water, known as condensation nuclei. The origin of these condensation nuclei is mostly in dust from the soil surface, salt particles formed by the evaporation of sea spray and industrial smoke. There are always sufficient nuclei present in the lower atmosphere for condensation to occur if the air is cooled to saturation, and in smoky industrial areas before the saturation point is reached.

It is further found that while impure water freezes in the laboratory at $0°C$, water droplets exist in the atmosphere at temperatures considerably below $0°C$. Liquid water at a temperature below $0°C$ is said to

exist in the supercooled state, and nearly all clouds which extend above the $0°C$ isotherm contain supercooled droplets at some stage in their history. In the atmosphere supercooled water droplets exist at temperatures down to $-40°C$, so it is only at very high levels in the troposphere that pure ice-crystal clouds, such as cirrus, are found. Clouds at temperatures between $0°C$ and $-40°C$ are observed to contain mostly supercooled water droplets with just the occasional ice crystal.

Clouds are essentially colloidal suspensions of minute liquid water drops and ice crystals. The water drops are extremely small being of the order of 1–100 μm in diameter, while the typical raindrop has a typical diameter of 1 mm (1 000 μm = 1 mm). The basic problem is therefore to explain the growth of cloud droplets to such a size that they can fall against the rising air currents and form rain. There are two main theories of this process, one known as the ice-crystal process and the other as the coalescence process.

The ice-crystal process was first suggested by Bergeron and is sometimes called the Bergeron process. He maintained that if a sufficient number of ice crystals in close proximity to water droplets are present in cloud, that there will be a movement of water-vapour from the droplets to the ice crystals since the saturation vapour pressure over ice is less than that over water. This difference in vapour pressure is largest in the $-10°$ to $-20°C$ range. The ice crystals then grow as snowflakes and as they fall they pick up more water by the accretion of small cloud droplets. If the temperature at ground level is at or below $0°C$, the resulting precipitation is in the form of snow, but if the snowflakes fall through warmer air in the lower atmosphere they melt and the resulting precipitation is in the form of rain. This process is a major mechanism for the production of precipitation in middle latitudes and accounts for both rain and snow, and the observed change from rainfall to snowfall with increasing altitude in mountainous regions.

In the tropics heavy rain is often observed to fall from clouds which do not reach the $0°C$ isotherm, which may be as high as 4 or 5 km. Under these conditions the rain cannot possibly be caused by the ice-crystal process because the cloud contains no ice, being completely above $0°C$ in temperature. Such observations have led to the development of the coalescence theory, which is not competitive but is complementary to the Bergeron idea. If the cloud contains some water droplets which are appreciably larger than the great majority of droplets, the slower rate of rise of such large drops in a cloud updraught leads to collisions and, in some cases, coalescence with the smaller droplets. Factors which promote this process are appreciable cloud depth and updraught speed which permit growth by accretion to a size sufficient to ensure that the drop will not evaporate at the top of the cloud but will fall back through the cloud, growing further by collision with small droplets and eventually reaching the ground as rain. The original large droplets probably form on extremely hygroscopic salt particles which in turn originate by the evaporation of spray from the sea surface. Evidence exists that the coalescence process is predominant at temperatures higher than $-12°C$, while the ice-crystal process is more important at lower temperatures.

Tropical rainfall will therefore form mostly by the coalescence process, and it is further observed to be mostly convective in origin. The convective clouds are organized into weather systems on various scales, and a characteristic feature of these weather systems is marked convergence of air in the lower atmosphere. The inflowing air provides a continual supply of moisture to the weather system and thus allows for the observed rainfalls. Warm moist equatorial air can hold in the form of water-vapour the equivalent of about 10 cm of rainfall, but an average equatorial shower will produce rainfall at the rate of 10 cm per hour for several hours, thus implying the need for a continual flow of water-vapour into the system. It is observed in many storms that the amount of rainfall produced is limited by the inflow of moist air, since if the inflowing air is dry the rainfall is small, while if it is very moist the rainfall is extremely heavy. Certainly for the production of continuous intense precipitation a strong inflow of very moist air into the storm is an essential condition in all parts of the world.

Evaporation and Evapotranspiration

At a water surface there is an exchange of water molecules to and fro between the water and the atmosphere; if more molecules are leaving the water surface to increase the water-vapour in the atmosphere, then evaporation is said to be taking place, the reverse process being known as condensation. Thus evaporation is the change of liquid water or ice to

water-vapour, and it proceeds continuously from the earth's free water surfaces, soil, snow and ice-fields. Evaporation can also remove water from plants and soil as well as from water bodies, but in this case the water loss is strongly influenced by the rate at which water can be supplied to the evaporating surface. The process by which liquid water contained in the soil is extracted by plant roots, passed upwards through the plant, and discharged as water-vapour to the atmosphere is known as transpiration. This takes place mainly through small openings in the plant leaves called stomata, which exist primarily to absorb carbon dioxide.

Evapotranspiration is the combined process of evaporation from the earth's surface and transpiration from vegetation. It depends on both the meteorological conditions and the amount of water in the soil, since evapotranspiration will be restricted by both a saturated atmosphere or a dry soil. So potential evapotranspiration is the maximum amount of water-vapour that can be added to the atmosphere under the given meteorological conditions from a surface covered by green vegetation with no lack of available water. Both evaporation and evapotranspiration are closely related to the radiation received, and may be considered as part of the energy balance discussed in Section B.

Direct observation of evaporation may be made by use of evaporation pans, where the water loss is determined on successive days by means of a micrometer hook gauge, any observed rainfall being added to the pan reading. Evaporation pans come in a variety of shapes and sizes. The most widely used is the U.S. Weather Bureau Class A pan, which is circular, 4 feet (1·22 m) in diameter, 10 inches (25·4 cm) in depth and is mounted near the ground on supports permitting a free flow of air around the pan. In contrast, the Standard British Pan is square with sides 6 feet (1·83 m) long, a depth of 2 feet (0·57 m), and is sunk into the ground with the rim of the pan projecting 3 inches (7·6 cm) above the surrounding ground. The relatively small area of a pan in comparison to a lake or a pond allows greater amounts of sensible heat from the atmosphere to be absorbed by the water in the pan through the sides and bottom. Thus pan evaporation is usually higher than that from a nearby large lake, and a correction coefficient has to be used to reduce the evaporation to that which would take place from a large free-water surface. The extraction of sensible heat by the pan is largest when the atmosphere is hot and dry and least when it is cool and humid, and so the largest pan coefficients are found in semi-arid climates. On average, the U.S. Weather Bureau Class A Pan has a coefficient of about $0·75$ while the British Pan is about $0·92$.

Evaporation pan readings, though of interest for reservoir construction, are not a measurement of potential evapotranspiration from vegetated surfaces, mainly because the energy balance of these surfaces is markedly different from that of free-water surface. The albedo of water is about $0·1$, depending on the altitude of the sun, but that of vegetation can vary from between $0·1$ and $0·3$, with a mean of about $0·25$.

The most practicable direct method of measuring natural evapotranspiration is by means of a lysimeter, the most accurate of which belong to the 'weighing monolith' type. Such an instrument consists of a block of soil, with plants growing in it, isolated from its surroundings by a casing. The whole block is weighed continually and it is assumed that short-term changes in weight are due either to the addition of moisture by precipitation or the loss of moisture by evapotranspiration. Correction factors have to be applied to lysimeter values in much the same way as evaporation pan readings have to be corrected. This is particularly so if the lysimeter is sited in a semi-arid climate and is being kept perpetually moist by the addition of known amounts of water so as to give a measure of the potential evapotranspiration. Under these conditions it will form a green oasis in an arid environment, sensible heat will be drawn from the atmosphere and used for evapotranspiration in addition to the net radiation, resulting in extremely high rates. This is known as the oasis effect and can be observed in semi-arid areas or in regions which undergo a seasonal dry period. It becomes less marked as the size of the isolated green area increases, until beyond a certain size which depends on local climatic conditions, it will completely vanish.

An alternative to the actual measurement of evapotranspiration is to attempt to estimate values using climatological data. Assuming that the supply of moisture is non-limiting, evapotranspiration is largely a function of net radiation and the drying power of the air. The former is important because evapotranspiration is an energy-consuming change, while the latter depends very largely on the relative humidity of the air and on the wind speed, since the lower the relative humidity and the stronger the wind, the greater will be the evapotranspiration. No evapotranspiration can take place into saturated stagnant air, and evapotranspiration is unlikely if the net radiation is negative. Evaporation can take place from small objects at night when the net radiation is negative, but heat is drawn from the air or surrounding objects to enable this to happen. This heat supply is limited, so while it will serve to dry wet clothes at night, it will not lead to significant evapotranspiration from a whole landscape. Thus potential evapotranspiration, which is being considered now because the moisture supply is not limited, is very strongly correlated with net radiation.

A particularly strong correlation exists for moist surfaces between monthly means of potential evapotranspiration and net radiation, and so also between monthly potential evapotranspiration and any variable which also happens to be strongly correlated with net radiation. Such a variable is monthly mean temperature, which is the basis for an empirical approach suggested by Thornthwaite (1948). In this, potential evapotranspiration is given by:

$$E_t = (h/12)(n/30)F(\overline{T})$$

where E_t is the potential evapotranspiration measured in cm/month,

h is the duration of daylight in hours,

n is the number of days in the month,

and $F(\overline{T})$ is a complicated function of the mean monthly temperature, T, and of a weighted annual mean temperature.

For values of $\overline{T}$ below $0°C$ it is assumed that the potential evapotranspiration is zero, while for values of $\overline{T}$ equal to $26·5°C$ it becomes $13·5$ cm/month in all climatic regions. Estimates of potential evapotranspiration are based therefore only on astronomical variables and on mean temperatures, since the factors $(h/12)$ and $(n/30)$ only correct the equation for season, latitude, and number of days in a month.

A fundamental weakness of Thornthwaite's approach is that net radiation is only considered implicitly through temperature, and so it is not suitable for time periods much shorter than one month. Penman (1948) has suggested an equation of the following form:

$$E_t = F\left(R_N, \bar{u}, e_s, e_a, \frac{de_s}{dt}\right)$$

TABLE 7.1
POTENTIAL EVAPOTRANSPIRATION VALUES AT SELECTED PENINSULAR MALAYSIAN SITES
(mm/day)

	Jan.	Feb.	Mar.	Apr.	May	June	July	Aug.	Sept.	Oct.	Nov.	Dec.
Alor Star	4·7	5·1	5·7	5·3	4·7	4·6	4·7	4·6	4·4	4·1	3·9	4·2
Kota Bharu	3·8	4·7	5·7	5·5	5·2	4·9	4·9	4·8	4·8	4·3	3·8	3·6
Penang	5·2	5·3	5·4	5·2	4·9	4·8	4·7	4·8	4·9	4·5	4·0	4·6
Kuala Trengganu	4·1	4·6	5·4	5·6	5·2	5·0	5·0	4·8	5·1	4·5	4·0	3·8
Ipoh	4·4	4·9	5·1	4·9	4·9	4·8	4·5	4·5	4·3	4·8	4·1	4·1
Sitiawan	3·9	4·5	4·7	4·7	4·6	4·7	4·6	4·6	4·5	4·4	4·0	3·8
Kuantan	3·4	4·2	4·8	4·8	4·6	4·3	4·3	4·2	4·4	4·0	3·6	3·3
Mersing	4·1	5·0	5·7	5·4	5·1	4·7	4·3	4·4	4·7	4·7	4·2	3·9

Source. (After J.G. Lockwood (1974). *World Climatology, An Environmental Approach.* (Edward Arnold, London).)

where R_N is the net radiation,

$\bar{u}$ is the mean wind speed;

e_s, e_a, and T are the saturated and actual vapour pressures and the temperature at screen level respectively,

$\frac{de_s}{dt}$ is the slope of the temperature—saturated vapour pressure curve at temperature T.

Normally values of net radiation are not available for use in Penman's equation, but he has suggested an empirical formula for calculating the net radiation from climatological observations including the number of hours of bright sunshine. Thus the method can be applied to data from climatological stations in many parts of the world, and gives very good estimates of · potential evapotranspiration values for periods longer than about ten days. Potential evapotranspiration values at selected sites in Malaysia are shown in Table 7.1. They are relatively uniform throughout the year and this reflects the very small seasonal changes in net radiation.

Interception

When rain reaches the surface it will normally first come into contact with vegetation, and it will only reach the soil surface after it has penetrated through the foliage. The processes involved are easily studied by sheltering under a dense tree during a rain shower. In the early stages of the storm the rain is absorbed by the tree canopy and the result is that the ground below the tree remains dry, but as the shower proceeds water starts to drip off the leaves and trickle down the trunk and the ground below the tree starts to get wet. Once the storm has ceased, the drips will continue for a short time until the surplus water has fallen from the tree and the water remaining in the tree will then evaporate into the atmosphere.

The above observations illustrate the three main components of interception, and these are shown in Figure 7.4. The first is the interception loss which is the water retained by the leaves and later evaporated away; the second is throughfall, which is water dripping through and from the leaves to the ground surface; and lastly stemflow, which is water that trickles along twigs and branches and finally down the main trunk to the ground surface.

The interception loss may be regarded as a primary water loss because it represents water which never enters the soil and the amount of loss depends on a variety of factors, the most important being the ability of the vegetation cover to collect and retain rainfall called the interception capacity of the vegetation. Even during rainfall, a considerable amount of water may be lost by evaporation from leaf surfaces, so that even when the initial interception capacity has been filled, there is some further, fairly constant retention of rainfall to make good this evaporation loss. Meteorological conditions will therefore partly determine the interception loss, and it could be large if the rainfall were light and the evaporation rate

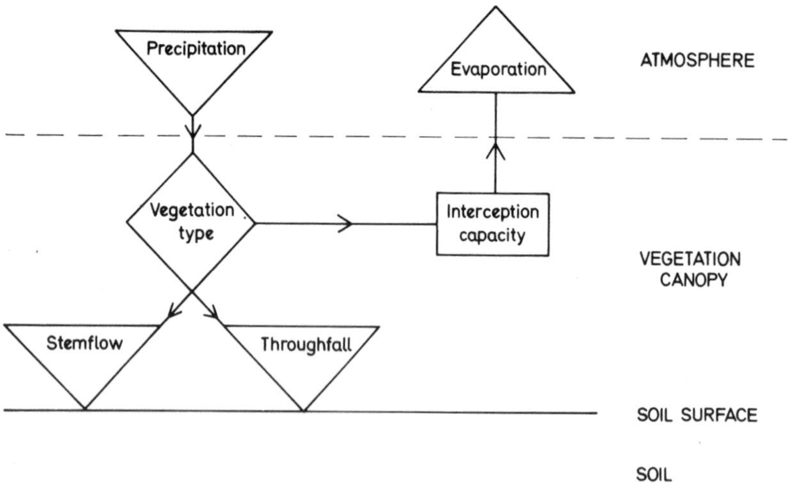

Figure 7.4 Interception depicted as a simple cascading system. Precipitation is the only input, while evaporation, stemflow and throughfall form the outputs. The interception capacity of the vegetation represents a store, while the vegetation type is a regulator.

high. This could occur in particular on a breezy day with light showers interspersed by sunny periods, since under these conditions all the shower precipitation could form interception loss and be evaporated, resulting in no water reaching the soil surface.

Interception loss by trees or tall vegetation may be measured by installing a randomly-spaced network of rain gauges in the vegetated area and comparing the catch with that measured in the open. Interception loss is the difference between the sets of observations after adjustment for stemflow, which may be measured by small collectors placed tightly around the trunks. Interception loss can be estimated by equations of the form:

$$I = C + et$$

where I is the interception loss for a given storm, C is the interception capacity of the vegetation cover, e is the average rate of evaporation during the storm, and t is the duration of the storm.

Observed interception loss values vary widely and depend not only on the vegetation type but also on the intensity of the storm. Indeed, it has been claimed that interception loss values for heavy rainfalls in tropical forest are less than those for light falls. Observations in tropical rain forests in Peninsular Malaysia suggest that with a total annual rainfall of about 250 cm, the interception loss amounts to

TABLE 7.2

STEMFLOW AND INTERCEPTION IN TROPICAL RAIN FOREST

A. *Stemflow. Average stemflow for storm classes (mm)*

Storm size	0·0—5·0	5·1—10·0	10·1—15·0	15·1—20·0	20·1—30·0	30·1—40·0	40·1—65·0
Stemflow	0·0	0·0	0·1	0·2	0·4	0·5	0·9

B. *Gross Rainfall and Interception*

Gross rainfall (mm)	1	2·5	5·0	7·5	10·0	15·0	20·0	30·0	40·0
Interception (mm)	0·7	0·9	1·2	1·5	1·8	2·4	3·0	4·2	5·4
Interception (per cent)	70·0	36·0	24·0	20·0	18·0	16·0	15·0	14·0	13·5

Source. (After I.J. Jackson (1971). 'Problems of throughfall and interception assessment under tropical forest'. *Journal of Hydrology*, 12, pp. 234-54.)

between 45 and 50 cm. Most of the rainfall seems to reach the soil surface in the forest by the process of throughfall, stemflow only being observed in the most intense storms. Stemflow has similarly been found to be small and even zero for many rainfalls in the tropical forest of Tanzania, and this is illustrated in Table 7.2. The percentage of the storm rainfall which eventually forms interception loss varies with the total storm rainfall, for in tropical forests over 70 per cent of the rainfall from storms with falls of 1 mm forms interception loss, while the corresponding value for falls of 40 mm is 13·5 per cent.

Infiltration

The flow or movement of water from the surface of the ground through pores and openings into the soil mass is known as infiltration. Gravity causes water to flow through the larger openings in the soil in appreciable quantities, but capillary forces can also be important since they disperse the water through smaller pores along the moisture gradient, although the movement is relatively slow and the quantity of water in motion is small. The movement of water through the soil mass to the water-table is referred to as percolation, which may therefore be viewed as a complementary process to that of infiltration.

Infiltration rates are very much dependent upon the physical state and characteristics of the soil. Soils which are coarse-textured such as sands have large pore spaces and the water movement is fairly rapid, while in fine-textured soils with small pore spaces it is very much slower. The nature of the vegetal cover

and the slope of the land also affect the rate of infiltration. Dense vegetal cover such as grass or forest tends to promote high infiltration rates since the dense rooting systems provide many passages into the soil.

Obviously, infiltration rates will vary from site to site, but nevertheless there will be at any particular site a maximum rate at which water can enter unsaturated soil, termed the infiltration capacity. If the rainfall intensity (Figure 7.5) is less than the infiltration capacity of the soil, then there will be no surface run-off and the infiltration rate will be equal to the rate at which water is supplied by the rainfall. In contrast, if the infiltration capacity is less than the rainfall intensity, then the infiltration rate will equal the infiltration capacity and the surplus water will form surface run-off. If the soil becomes saturated then all the additional rainfall will form run-off.

During a rainstorm a number of changes take place in the physical properties of the soil which affect the infiltration capacity. These physical changes result from the packing of the soil surface by the rain, the swelling of the soil to close openings through which infiltration would take place by gravity, the washing of fine and colloidal materials into the soil surface openings, and the general increase in soil moisture. The net result is to reduce the infiltration capacity progressively during a rainstorm until it reaches a steady minimum value. Horton (1939) suggested that the infiltration capacity can be represented by the equation:

$$f_p = f_c + (f_o - f_c)\exp(-kt);$$

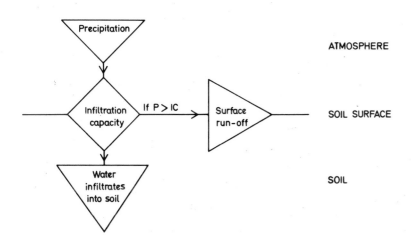

ATMOSPHERE

SOIL SURFACE

SOIL

Figure 7.5 Infiltration. The movement of water at the surface of unsaturated soil is regulated by the infiltration capacity of the upper layers of the soil. When it is raining, water usually moves downwards into the soil, but if the intensity of the rainfall exceeds the infiltration capacity of the soil then the excess water forms surface run-off. If the soil becomes saturated then all the additional rainfall will form run-off.

where f_p is the infiltration capacity at time t from the beginning of the rain, f_o is the initial infiltration capacity, f_c is the minimum constant infiltration capacity, and k is a constant. It is found in practice that as the high initial infiltration rate only lasts for a few minutes, it is therefore of minor importance in reducing floods from rainstorms of several hours' duration.

In some soils the infiltration capacity of the A horizon is very high, but this is underlain by a very clayey B horizon in which percolation rates are extremely low. In this case rain-water moves rapidly through the top horizon but soon becomes ponded above the B horizon where rates of water movement are very slow. This can result in the top horizons of the soil becoming saturated and then any additional rainfall, regardless of its intensity, will form run-off.

Surface run-off can result therefore from either the rainfall intensity exceeding the infiltration capacity of the soil or the soil becoming saturated because of the drainage being impeded at some depth below the surface horizon.

Infiltration capacities may be measured by use of portable infiltrometers or by an analysis of river-flow after rainfall. There are various types of portable infiltrometers, and one such instrument used in Peninsular Malaysia has been described by Eyles (1967). It consisted of two constant head tanks with outlet valves to control flow to a sprinkler head, which was used to apply water from the large tank to the interior of a steel cylinder (area 1 sq ft $(0.092\ m^2)$) driven 6 inches (15 cm) into the ground. When the intake rate had settled down to a constant value, the small tank was used to adjust the rate of application so that it equalled the rate of absorption. As the rate of flow from the tanks was known from calibrations carried out in the laboratory, it was possible to measure the infiltration capacity by noting the rate at which the water-level dropped in the small tank. Eyles comments that a number of errors are inherent in this method; these include the subjectivity of equating the rates of application and intake, the disturbance of the soil which accompanies the placing of the cylinder, and the lateral escape of air and water at the base of the cylinder. However, the estimates which the infiltrometer provides on the relative infiltration capacities of various soil and land-use types can be very useful in a variety of studies. It is possible to improve on the portable infiltrometer by choosing

small 'homogeneous' drainage basins and carefully measuring precipitation, evaporation and surface run-off, for it is then possible to compute the average infiltration.

Since infiltration capacities are extremely variable, it is difficult to quote typical values. Using the instrument just described on the Malacca soil series in a rubber estate in Peninsular Malaysia, Eyles found that the infiltration capacity varied from 1·5 to 42 cm per hour with a mean of 14·7. These observations can be compared with those from the Batu Anam soil series on the same estate where infiltration capacity varied from 1·0 to 1·4 with a mean of 1·2 cm per hour.

Soil Moisture

If gravity were the only force to which water in the soil were subject, the soil would drain completely dry after rainfall, and the only water found would be that below the water-table. However, under natural conditions, the soil always contains some moisture, even after very long dry periods lasting several months, indicating that the forces which hold moisture in the soil are extremely strong.

The nature of soil moisture storage may be investigated by considering the gradual drying of a soil after prolonged rainfall. At first the soil will be completely saturated and there will be rapid downward drainage of excess water under gravitational action. After a day or two a stage will be reached when the rate of drainage has become much slower and the bulk of the water is held in the soil by capillary forces since the soil is now at field capacity. These capillary forces arise from surface tension effects, which result because molecules on the water surface are attracted mainly by molecules within the water body and only infinitesimally by the water-vapour molecules in the air. There is therefore a certain amount of free energy, known as surface tension, at the liquid surface. This may be illustrated by placing a very fine capillary tube in water; it is observed that the surface tension force causes the water to rise up the inside of the tube to a height which is inversely proportional to the diameter of the tube. In the soil thin films of water are held around soil particles and in small spaces between soil particles. The force holding the water is greatest in the smallest spaces and in the thin films around the smallest parti-

TABLE 7.3
SOIL MOISTURE TENSIONS

Soil Moisture Tension (atmospheres)	State of Soil
1/1 000	Saturated with standing water. Drainage of water under gravity.
1/3	Field capacity. Excess water has drained away. Water loss now partly by drainage and partly by evapotranspiration.
15	Wilting point. Significant drainage and evapotranspiration have now ceased.
10 000	Soil almost completely dry.

cles, and since these capillary forces have to be overcome to extract water from the soil, it would appear that the smaller the soil water content, the greater the force with which the remaining water will be held in the soil. One method of specifying soil moisture content is to state the suction which has to be applied to just remove water from the soil; this is known as the soil moisture tension (Table 7.3).

As the soil continues to dry, both by evapotranspiration and slow drainage, the soil moisture tension increases, and therefore the resistance to further moisture loss increases. One result of this is that actual evapotranspiration will fall below the estimated potential values. Plants extract moisture from the soil into their rooting systems against the prevailing soil moisture tension by the process of osmosis, but as the drying of the soil proceeds a point is reached where the soil moisture tension has increased to a value such that the plants are unable to extract water from the soil and therefore wilt and die. This is known as the wilting point and for average plants corresponds to a soil moisture tension of about 15 atmospheres. When soil is at field capacity, that is to say after excess water has drained away following heavy rain, the soil moisture tension is between 0·1 and 0·4 atmospheres, and the amount of water normally available for evapotranspiration from plants is represented by the difference in the water stored per unit volume at field capacity and at the wilting point, this being known as the available water content of the soil.

At the permanent wilting point some of the soil water is held by capillary action, but a large amount is also held by absorption on individual soil particles, whereby water molecules are attracted and adhere strongly to the surface of solid particles without any chemical combination taking place. This water, which is the last to be lost from the soil, is called hygroscopic water and normally forms only a minute part of the total soil moisture storage. Even at the permanent wilting point the soil can still lose moisture, and the soil moisture tension under severe drought conditions may rise to 40 or 50 atmospheres.

The measurement of the water content of soil is not easy, and in practice it is often calculated as part of the water-balance, which is considered later in this chapter. The most accurate method of measuring the water content of soil involves determining the weight loss from oven-dried samples. Each field sample is weighted before and after being dried at a temperature of 105°C; the moisture content is then normally regarded as the ratio of the weight of water lost during drying, to the weight of the dry soil, expressed as a percentage. At a temperature of 105°C most of the hygroscopic water will remain in the soil, and the water loss measured will be that which could occur from normal evapotranspiration. At higher drying temperatures the hygroscopic water will be driven off and the soil will become completely dry. If an oven is not available, the soil sample can be allowed to dry naturally after the first weighing, but this may take several days. There are various methods of measuring soil moisture directly in the field but they all involve complex instrumentation, and therefore reference should be made to hydrological textbooks for further details.

Groundwater

Precipitation which infiltrates into the soil and penetrates to the underlying strata is called groundwater, and the water-bearing strata are called aquifers which often consist of consolidated sandstones, or unconsolidated sands, gravels and glacial drifts. Water in the pores of an aquifer is subject to gravitational forces and tends to flow downwards through the pores of the material until it reaches a zone where the aquifer is saturated, this surface of saturation being referred to as the groundwater table or the phreatic surface. Because of the increasing compression of the rocks with depth, their porosity decreases with increasing depth and this is reflected by a corresponding decrease in water content. Water does not sink to great depths to be lost to the hydrological cycle but instead the water below the phreatic surface flows slowly towards the nearest free-water surface such as a river, lake or even the sea.

If the geology of the rocks through which it passes is complex, various complications will be introduced into the groundwater flow. For example, an impermeable layer may underlie an aquifer, and if they both outcrop at the surface, there will be a zone of seepage or a line of springs along the bottom of the aquifer. Sometimes a small area of impermeable rock may exist in a large aquifer and this may give rise to a perched water-table which is above the true phreatic surface. It is equally possible for a groundwater aquifer to be overlain by impermeable material and so be under pressure; such an aquifer, which will be fed from a distance, is called a confined aquifer, and the surface to which the water would rise if allowed, is called the piezometric surface. If the piezometric surface is above ground level, water will flow out of a well bored from the surface into the aquifer, and this well is known as an artesian well.

The amount of groundwater stored in a saturated material depends upon the porosity of the material, which is given by:

$$\text{porosity} = \frac{\text{total volume of voids}}{\text{total volume of material}}$$

All voids in the material are involved in the concept of porosity including spaces between individual particles and also joints, bedding planes and fractures.

Groundwater moves extremely slowly and its rate of movement is partly controlled by the permeability, which is a function of the porosity, structure and geological history of the material, and which may be measured in terms of the permeability coefficient, expressed in metres per day or feet per day. Since the movement of groundwater is subject to the influence of gravity in much the same way as surface water, it flows towards the point where the phreatic surface is lowest at a velocity which is proportional to the slope of the phreatic surface. If a material is considered in which there is complete uniformity with a steady state of groundwater flow, then the rate of flow per unit area (Q) may be stated in the terms of D'Arcy's law:

$$Q = Ki$$

where K is the permeability coefficient, and i is the hydraulic gradient (slope of the phreatic surface).

Total water storage in an aquifer represents the interaction between groundwater discharge and groundwater recharge:

$$\text{Water stored} = \frac{\text{Recharge of}}{\text{groundwater}} - \frac{\text{Discharge of}}{\text{groundwater}}$$

The loss of water from storage is usually continuous, while the recharge is often discontinuous and sometimes restricted to certain seasons of the year. Even under rainless conditions, water flows out of groundwater storage into rivers and lakes, and it is then found that the amount of water stored decays in an expotential manner. The main components of groundwater recharge may be listed as follows:

(i) infiltration and percolation from the soil surface;

(ii) influent seepage through the banks and beds of surface water bodies such as ditches, rivers and lakes;

(iii) underground leakage of water from adjacent aquifers;

(iv) in certain cases the groundwater may be recharged artificially.

Since the level of the phreatic surface will rise and fall as the amount of groundwater storage increases or decreases, it may be taken as a measure of the amount of groundwater storage. The level of the phreatic surface may be measured by noting the water-level in unused wells, where it is observed to be highest in the wet season and lowest in the dry season.

Water-balance

The various components of the hydrological cycle may be linked together in terms of a cascading system, that is to say a system composed of a chain of sub-systems, often characterized by thresholds, having both spatial magnitude and geographical location, which are dynamically linked by a cascade of mass or energy. From the previous discussion it is clear that the hydrological cycle can be divided into a number of sub-systems, each of which has inputs from other sub-systems, some form of storage and internal regulator, and outputs to other sub-systems.

A whole series of sub-systems can be recognized,

and together these form the hydrological cycle. The atmosphere forms the base of one such sub-system, the input being evaporation from the oceans and moist land surfaces while the output is in the form of precipitation. Storage takes the form of atmospheric water-vapour and the release of water in the form of precipitation from this storage is governed by the meteorological processes within the atmosphere. The output from the atmospheric sub-system forms the input into the next sub-system, which is the vegetation/surface sub-system. Of the precipitation falling on the surface, some is intercepted and stored either on the vegetation or on the surface, some evaporates, some forms surface run-off and some infiltrates into the soil. In this case there are several outputs from the sub-system, of which the two most important are surface run-off and infiltration, the ratio between these being controlled by the infiltration capacity of the soil.

Soil forms yet another sub-system in which the input is infiltrating water from above and the output is either percolation to the water-table, flow down-slope parallel to the surface, or evapotranspiration. The soil moisture is the storage component in this case, and it also acts as a regulator in that only moisture near or in excess of the field capacity will readily form outputs. Various other sub-systems may also be identified, such as the groundwater sub-system, the river channel sub-system, the ocean sub-system, etc.

It is possible to take any of the sub-systems and produce a sort of balance-sheet which shows at any time the magnitudes of the inputs, outputs and storage capacity of the system. In this particular case the balance-sheet so produced is termed the water-balance of the particular sub-system under consideration. Water-balances may be produced for any of the sub-systems or indeed for combinations of the sub-systems, but it is most usual to produce them for the soil moisture sub-system.

Water-balance calculations for the soil are often made with the assumptions that the only input is precipitation and that the outputs are just evapotranspiration and horizontal throughflow or run-off. This is reasonable since the vertical movement of water downwards into the deep rocks is usually very small when compared with evapotranspiration and run-off. While the surface is saturated, evapotranspiration is at the potential rate, but as soon as it becomes dry evapotranspiration falls below the potential rate. The upward movement of water in soil is extremely slow compared with the potential evapotranspiration rates observed on a warm sunny day. Thus under conditions of moderate or high evapotranspiration rates, the very top layers of even saturated soils will rapidly become dry, and this has been observed to occur after only about 0·2 cm of water had been evaporated. In water-balance calculations it is assumed that the soil surface is covered by freely transpiring vegetation

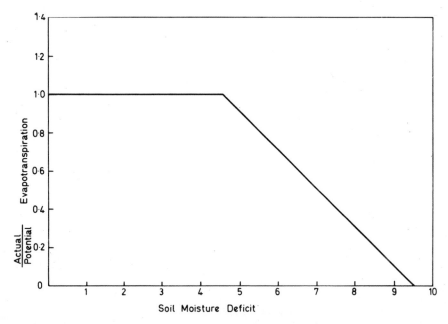

Figure 7.6 The ratio of actual to potential evapotranspiration under drying conditions. The ratio of actual to potential evapotranspiration as a vegetation-covered soil dries with a resulting increase in the soil moisture deficit. The potential evapotranspiration is assumed to be at a constant rate throughout.

TABLE 7.4

ESTIMATIONS OF WATER AVAILABLE FOR EVAPOTRANSPIRATION IN SOME
TYPICAL TROPICAL RAIN FOREST SITUATIONS

| *Forest Type* | *Soil Type* | *Water Available in* | |
		(a) *Soil* (mm)	(b) *Crop* (mm)
Mixed dipterocarp forest	Red-yellow Podzols	180	20
,,	Lateritic	160	25
Kerangas forest	Deep humus Podzol	552	25
,,	Shallow humus Podzol	60	5

which is extracting water from some depth of soil. The actual soil moisture available to plants for transpiration depends on the rooting depth of the vegetation and the soil characteristics, and for small field crops on an average soil it is usually equivalent to about 10 cm of rainfall. The moisture available to trees is greater because their roots penetrate more deeply into the soil, and in parts of the tropical rain forest of Sarawak can reach 55 cm. Evapotranspiration is therefore able to continue at the full potential rate for a considerable period of time from a field crop, but it would be severely restricted if the crop were cleared and a bare soil surface left. Observations (Figure 7.6) from a considerable variety of crops suggest that as the soil dries out, evapotranspiration continues at the potential rate until a certain limiting value of soil moisture is reached after which it falls below the potential rate. If the potential evapotranspiration rate remains constant, then the ratio of actual to potential decreases linearly as the soil moisture deficit increases past the limiting value. The soil moisture deficit is the amount of rainfall needed to bring the soil to field capacity at any given time. The exact value of the soil moisture deficit at which the vegetation cover ceases to act as a saturated surface (i.e. actual and potential evapotranspiration cease to be equal) is not constant but varies with both the vegetation and soil type. This is well illustrated by Table 7.4 which indicates the water available for evapotranspiration in some typical tropical rain forest situations.

The usual approach to soil moisture balance calculations is to subtract the potential evapotranspiration from the rainfall. If the rainfall exceeds the potential evapotranspiration then the excess moisture goes towards changing the soil to field capacity and after this is attained, to forming run-off. When potential evapotranspiration exceeds rainfall, the deficit is made up by the evaporation of water stored in the soil, and in the very simple models it is assumed that this will continue until all the available soil moisture is evaporated. If at this stage potential evapotranspiration still exceeds the rainfall, the water deficit cannot be made up by the evaporation of soil moisture and the actual evapotranspiration becomes equal to the rainfall. Sample water-balances, calculated by this method, are given in Table 7.5 for Canton and Hong Kong.

The fall of actual evapotranspiration rates below potential values under drying soil conditions can be taken into account by using multilayer soil models. In these models the layers may represent actual physical layers in the soil or they may represent stages in the drying of the soil. Often it is assumed that the top layer or first stage receives all the incoming rainfall and that evapotranspiration takes place at the full potential rate. Excess water in the top layer either runs off or infiltrates into the second layer. When under drying conditions the moisture content of the top layer or first stage is exhausted, it is assumed that evapotranspiration takes place from the second layer or stage at some rate below the potential. Since under constant potential evapotranspiration the relationship between actual and potential varies in a linear manner with soil moisture deficit, the assumption is often made for the second layer or stage that:

$$\begin{array}{l} \text{actual evapo-} \\ \text{transpiration} \end{array} = \text{potential evapotranspiration}$$

$$x \; \frac{\text{actual soil moisture content of second layer}}{\text{saturated soil moisture content of second layer}}$$

A soil moisture balance for Alor Star from October 1955 to September 1956 computed by this method is shown in Figure 7.7. It is assumed that the first layer had a storage capacity of 25 mm and the second one of 125 mm, making the available water content of the soil 150 mm. The potential evapotranspiration for the period was 1 609 mm, while this model suggests that the actual evapotranspiration for the same period was 1 447 mm.

Long-term studies of the water-balance of forest catchments do not exist for Malaysia, but a number of short-term studies have been made which can be

TABLE 7.5
AVERAGE ANNUAL WATER-BALANCE
Water-Balance at Canton (mm)

Month	Rainfall	P.E.	A.E.	Soil Moisture	Run-off
Jan.	27·1	56·7	27·1	—	—
Feb.	65·0	54·9	54·9	10·1	—
Mar.	100·7	67·2	67·2	43·6	—
Apr.	184·7	81·6	81·6	100·0	46·7
May	256·1	119·1	119·1	100·0	137·0
June	291·5	126·0	126·0	100·0	165·5
July	264·4	157·2	157·2	100·0	107·2
Aug.	248·5	149·4	149·4	100·0	99·1
Sept.	149·1	135·6	135·6	100·0	13·5
Oct.	48·5	115·8	115·8	32·7	—
Nov.	50·5	78·9	78·9	4·3	—
Dec.	34·1	62·4	38·4	—	—

Water-Balance at Hong Kong (mm)

Month	Rainfall	P.E.	A.E.	Soil Moisture	Run-off
Jan.	32·5	79·0	32·5	—	—
Feb.	45·0	76·0	45·0	—	—
Mar.	72·5	86·0	72·5	—	—
Apr.	135·0	97·0	97·0	38·0	—
May	287·5	128·0	128·0	100·0	97·5
June	387·5	127·0	127·0	100·0	260·5
July	375·0	146·0	146·0	100·0	229·0
Aug.	355·0	150·0	150·0	100·0	205·0
Sept.	252·5	138·0	138·0	100·0	114·5
Oct.	112·5	137·0	137·0	75·0	—
Nov.	42·5	113·0	113·0	5·0	—
Dec.	30·0	89·0	35·0	—	—

P.E. — Potential Evapotranspiration
A.E. — Actual Evapotranspiration
Soil Moisture calculated assuming a field capacity of 100 mm.

Source. (After J.G. Lockwood (1974). World Climatology, An Environmental Approach. (Edward Arnold, London).)

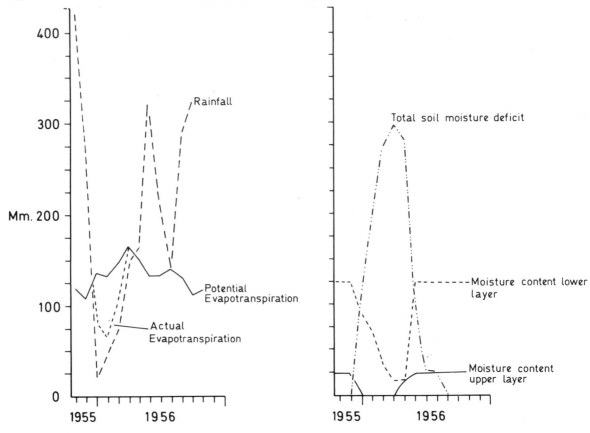

Figure 7.7 Water-Balance at Alor Star using a two-layer soil model. The upper layer has a total storage capacity of 25 mm while that of the lower layer is 125 mm. Rainfall and evapotranspiration take place directly from the upper layer, and excess water is transferred to the lower layer. When the upper layer is dry, evapotranspiration takes place from the lower layer according to the model suggested in Figure 7.6. All values are in rainfall equivalents.

Left. Rainfall and potential evapotranspiration and actual evapotranspiration calculated from the model.

Right. Total soil moisture deficit calculated assuming that evapotranspiration is always at the potential rate. Soil moisture curves for the upper and lower layers are also shown.

considered to give reasonably typical values. Thus Kenworthy (1970) studied the water-balance of a very small forest catchment on the north side of the Sungai Gombak, Central Range, Malay Peninsula. This particular catchment drained an area with an elevation between about 250 m and 580 m, and on the higher slopes the vegetation was primary dipterocarp forest, but on the lower slopes it had been disturbed by logging operations. Over a relatively short period of time, Kenworthy estimated that with an annual rainfall of 2 500 mm, the actual evapotranspiration was about 1 750 mm per year. He further estimated that 1 300 mm per year was used in transpiration, and 450 mm was intercepted by the canopy and lost again as direct evaporation; 100 mm was lost

as direct run-off over the soil surface, the remaining 650 mm being lost by flow through the soil and the deeper rocks.

Within individual years, run-off shows marked variations which largely reflect the annual march of rainfall. This is illustrated in Table 7.6 which lists rainfall, run-off, estimated potential evapotranspiration, and estimated soil moisture deficit for the Damansara catchment near Kuala Lumpur. Major types of land use within this catchment include 43 per cent rubber plantations and 36 per cent primary forest, and it is reasonably typical of agricultural areas near Kuala Lumpur. In the average year at Kuala Lumpur there are rainfall maxima around April and November, and a well-developed minimum in

TABLE 7.6
WATER-BALANCE OF DAMANSARA CATCHMENT, SELANGOR
(Units. Rainfall equivalent in mm)

Date	Rainfall	Estimated Potential Evapotranspiration	Measured Run-off	Estimated Soil-Moisture Deficit
1968				
Oct.	345	132	148	0
Nov.	200	120	158	0
Dec.	315	114	117	0
1969				
Jan.	248	117	168	0
Feb.	144	135	77	0
Mar.	211	141	131	0
Apr.	228	141	99	0
May	328	138	261	0
June	209	141	137	0
July	117	138	79	−21
Aug.	275	138	168	0
Sept.	125	135	70	−10
Oct.	356	132	186	0
Nov.	250	120	226	0
Dec.	211	114	124	0

Source. (Rainfall and run-off data from Goh Kim Chuan, 'A comparative study of the rainfall-runoff characteristics of a developed and a forested catchment'. Unpublished M.A. thesis, University of Malaya, 1972.)

July with a secondary minimum in January/February. Also, in the average year there is a small soil moisture deficit between June and August. All these features were present in 1969 and may be seen in Table 7.6. The observed run-off was least during the periods of soil moisture deficit and greatest during the periods of high rainfall. The relatively low run-off values in October 1969, probably reflect the large amounts of water drawn from soil and groundwater storage in the previous months, since this would have to be replaced before very large run-off totals could occur.

8 Rivers and River Systems

WHENEVER the rate of precipitation exceeds the infiltration capacity of the soil surface or the soil becomes saturated, then the excess water will accumulate upon the surface. At first, the excess water is stored in irregularities on the surface, but when their storage capacity is exceeded, surface run-off begins as a thin sheet-flow. The uneven nature of the surface soon concentrates the sheet-flow into channels, which further join in a tree-like structure which ensures that the flow immediately below each confluence exceeds that in either of the merging branches. Not only may water reach the main river by way of surface run-off and minor stream flow but also by movement just below the surface which is known as sub-surface flow, interflow or throughflow, and this can be the main form of near surface run-off when the infiltration capacity of the soil is not exceeded. Rivers do flow during dry weather when there is no surface run-off,

and in this case the water is supplied by the outflow of groundwater from the deep rocks. Whereas surface and sub-surface run-off is rapid and therefore causes sudden increases in the volume of flow of rivers, the groundwater component of the flow is relatively steady and provides the basic background flow of most river systems.

The water-balance of a small area of the earth's surface over a short period of time may be expressed by:

$$P + SF_{IN} + G_{IN} = SW + GW + E_t + SF_{OUT} + G_{OUT}$$

where P is the precipitation,

E_t is the evapotranspiration,
SW is the change in soil water storage,
GW is the change in groundwater storage,
SF_{IN} and SF_{OUT} are the water flows over the surface into and out of the area respectively,

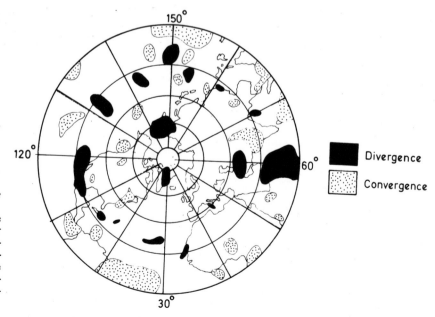

Figure 8.1 Average convergence and divergence of water vapour. Regions of marked convergence or divergence of water vapour are shaded. Convergence indicates that precipitation exceeds evapotranspiration, while divergence indicates that evapotranspiration exceeds precipitation.

and G_{IN} and G_{OUT} are the groundwater flows into and out of the area respectively.

If this equation is slightly rearranged it becomes:

$$P - E_t = SW + GW + (SF_{OUT} - SF_{IN}) + (G_{OUT} - G_{IN})$$

Since the soil water and groundwater storage capacities have finite maximum values which are usually small, there will be a flow of surface and groundwater out of the area if precipitation exceeds evaporation. As precipitation is a function of the atmospheric circulation and occurs in zones where there is a strong convergence of water-vapour (Figure 8.1), this suggests that the distribution of the major river systems is partly a result of the atmospheric circulation. This is reasonable to assume because the flow of rivers and the movement of water-vapour in the atmosphere are both parts of the hydrological cycle and therefore are clearly interrelated.

If precipitation exceeds evaporation then there will be a flow of water-vapour into the area to make up the deficit and similarly water-vapour will flow out of an area if evaporation exceeds precipitation. In the atmosphere, areas of general convergence of water-vapour are found near the equator indicating an excess of precipitation over evaporation due to the convergence of the trade winds from both hemispheres, while general divergence is found in the sub-tropical anticyclones. In the middle and high latitudes there are alternating areas of convergence and divergence. Two areas of strong convergence within the tropics are of special interest. The first which extends from south-western Arabia generally westward and southward across equatorial eastern and central Africa contains the headwaters of the Blue Nile, several tributaries of the White Nile, the upper parts of the Congo, and Ubangi rivers, and there are also several rivers flowing southward from this area through Kenya. The second water-vapour convergence zone which is over north-central India extends northward through Kashmir to the Pamirs and Altai Mountains west of Sinkiang. The heavy rains over India are well known and the Indian convergence zone covers the headwaters of such extensive river systems as the Indus, Ganges, Brahmaputra, Salween, Mekong and Yangtze. The water-vapour entering these equatorial convergence zones originates from the oceans under the sub-tropical anticyclones which are regions of water-vapour divergence.

The landscape is divided into regions which form natural run-off systems, and these form very convenient units for studying river systems. Under normal conditions the run-off from these drainage regions will flow into a river or lake. It is therefore possible to define the surface drainage area of a stream or lake as that area which is enclosed by a topographic divide such that direct surface run-off resulting from precipitation will drain by gravity into the lake or stream. If a whole river system is being considered, the drainage area will extend from the junction with the sea to the topographic divides, but if only part of a river system is under discussion, the drainage area will close at some point on the river. The term 'drainage basin' is commonly used in American literature to mean the area enclosed by the boundaries of the surface run-off system, while its British equivalent is called the 'catchment'.

So far, only surface drainage systems have been considered, but obviously there will be underground flow into the sea or lake, and therefore an underground drainage basin can also be defined. The boundaries of the surface catchment are determined by local relief and will be a natural ridgeline, but in contrast the underground catchment will be determined by the local geology and need not be coincident with the surface catchment.

Though there is, by definition, no export of water across the horizontal boundaries of the surface catchment, except at the lowest point, the system is open vertically since it is able to receive water in the form of precipitation and lose it via evapotranspiration. Catchments may therefore be considered as examples of open physical systems in that they exchange both mass and energy with their surroundings.

Hydrographs

The hydrograph describes the variation of river-flow with time. River-flow is measured in terms of the volume of water passing a particular section of the river in a given time, the most usual units being cubic metres per second (cumecs) or cubic feet per second (cusecs). So the hydrograph (Figure 8.2) is a curve with time along one axis and volume along the other.

Under dry weather conditions the river-flow consists completely of water contributed by the aquifers bordering the river. Since groundwater forms most of the continuing long-term flow of the river, it is said to form the so-called base flow, and under rainless conditions the river is said to be in base flow. Since aquifers discharge decreasing amounts of water with time, as the stored water decreases, the base flow slowly decays with time in an exponential manner (see Chapter 1). The base flow hydrograph can be represented very nearly by:

$$Q_t = Q_o e^{-\alpha t}$$

where Q_o is the discharge at the start of the period,

Q_t is the discharge at the end of time t,

α is the coefficient of the aquifers in the catchment,

and e is the base of natural logarithms.

This equation states that if the flow today is known at a given section on a river, it can be forecasted for any time in the future, assuming that there is no precipitation. It also explains how it is that many rivers manage to keep flowing even after exceedingly long dry spells. Decrease in river-flow is very rapid just after a flood peak following heavy rainfall, but the rate of decrease is not uniform but decreases itself with time, and the amount of decrease in flow is always less than the total flow at that instance in time, so though the river-flow may reach very small values it will take a very long time to vanish completely.

When precipitation occurs which is sufficiently intense or prolonged to produce surface or sub-surface flow, water will start to move relatively quickly across the landscape into the river and the hydrograph will start to rise. Infiltration, soil moisture and interception were discussed in the previous chapter, where it became clear that significant surface or sub-surface flow would occur when the rainfall intensity exceeded the infiltration capacity of the soil or when the surface layers of the soil became saturated. Thus heavy rain on a dry soil with a high infiltration capacity may lead to no significant run-off, while a light rain on a saturated soil could lead to almost complete run-off. Sub-surface flow in the top layers of the soil will probably occur first but its rate of movement is considerably slower than true surface run-off.

River hydrographs (Figure 8.2) during and following significant precipitation may be explained by considering the flow at a segment of the river, which forms the exit to the catchment above that particular point. Water will take some time to flow down the valley slopes and along the river to the catchment exit and obviously the further the water has to travel the longer it will take to reach the exit. The speed of travel will depend very much on the nature of the route followed, water in stream channels and overland flow moving much more quickly than water seeping through the soil. Now for the purpose of this discussion it is assumed that the intensity of the rainfall is uniform over the whole catchment and that the rain starts and finishes at the same time everywhere.

Figure 8.2 Flood Hydrograph. When the weather is dry for some time, river flow is maintained by water contributed from aquifers bordering the river. This is the base flow which decreases in an exponential manner. When it rains significantly, water moves across the landscape and into the river, causing the flow to increase. The first water to arrive is that which has fallen in or near the channel, then progressively more distant parts of the catchment contribute to the flow, causing it to increase, even if the rainfall is constant, until the whole catchment is contributing. When the rain stops, the flow decays as surface and sub-surface run-off decrease until eventually the river returns to the base-flow condition.

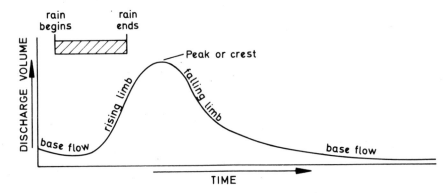

It also assumes that the soil moisture and infiltration capacity are uniform over the whole catchment.

If it has been dry for some considerable period of time before the rainfall, then the river-flow will consist completely of base flow, i.e. of groundwater which has seeped into the river. Once surface run-off reaches the river after the start of rainfall, the river-flow will consist of both surface run-off and base flow, and as the hydrograph rises the surface run-off will become the dominant part of the river-flow. Since seepage into and out of the aquifers in the catchment is slow, the base flow will remain steady or only increase slowly throughout the storm, and it may therefore be assumed that the short-term variations in river-flow are largely due to surface run-off.

Water to arrive first at the catchment exit will be that which has fallen nearest to it, since the travel time will be least. Indeed, rain which has fallen onto the river surface just upstream will be the first to arrive. As the storm proceeds, water from further and further up the catchment will reach the exit, causing the flow at the exit to increase. Looked at in another way, it can be said that the active catchment area contributing to the flow at the exit increases with time, thus causing the stream hydrograph to rise. Eventually the catchment which contributes to the flow at the exit and the topographic surface catchment coincide, run-off from the whole of the surface catchment reaches the exit, and the river-flow stops increasing. The time taken for water from the most distant parts of the catchment to reach the outlet at the lowest point is known as the time of concentration of the catchment and is a measure of its size. Thus even with rainfall of constant intensity, the stream hydrograph will rise until the time of concentration is exceeded, after which it will remain constant because the area contributing to the run-off is now fixed. Once the time of concentration is exceeded, the river-flow will only vary with variations in rainfall intensity.

Eventually the rainfall will cease and the reverse to the above process will take place, that is to say, the last surface run-off to reach the outlet will be from the furthest point of the catchment, and as a result the hydrograph will decay until it is being produced completely by base flow originating in the deep rocks. Because of the time taken for water to move across the landscape the stream hydrograph may not start to decay until some time after the rainfall has ceased, and indeed, if the time of concentration is not exceeded, the hydrograph may actually continue to rise for a time after the end of the rainstorm. The characteristics of the decaying hydrograph can therefore only depend on the physical features of the catchment.

There is an approximate climatological relationship between storm area and storm duration, since normally the larger the area of a storm, the longer the duration. Thus showers are of a short duration (15 min) and cover a small area (20 km²), while major cyclonic disturbances may last a whole day and cover half a country. Now water movement across catchments is slow and it is very slow compared with the duration of significant storms over the catchment. If a storm just covers a catchment, that is to say it produces rain over the whole of the catchment, then the duration of the storm is often found to be considerably less than the time of concentration of the catchment. Therefore under these circumstances the total area of the catchment will never contribute at one time to the hydrograph, which will be truncated before it is fully developed by the ceasing of the storm. If all storms were equally intense, this phenomenon would not be very obvious in small catchments, but normally the average intensity of rainfall increases as the average size and duration of the storm decreases. So the most marked hydrographs in any catchment will usually be produced by storms which are the same size or larger than the catchment and which also have durations longer than the time of concentration of the catchment. The best developed hydrographs will normally be produced by the smallest storms which satisfy the conditions just mentioned. Since storms rarely fulfil these conditions, many river hydrographs appear always to be a function of both storm intensity and duration. Real hydrographs are often of a complex nature because water moves at different rates across the landscape depending on its exact path. Thus water falling on the stream surface will give rise to a very rapid response in the hydrograph, but the response from water moving through the soil will be very much slower.

Storm characteristics and also the physical characteristics of the catchment are important in determining the shape of the stream hydrograph. It has already been shown that storm size, duration and intensity are significant. The catchment characteristics that are of interest include the following:

(i) Catchment area—the greater the area the larger the storm run-off.

(ii) Catchment shape—this very effectively controls the shape of the hydrograph.

(iii) Catchment slope—the more steeply the surface slopes the more rapid is the run-off, so that the time of concentration will be shorter and the flood peaks higher.

(iv) Surface conditions—such as land use (rubber plantations, tropical forest, grassland etc.).

(v) Sub-surface conditions—such as initial soil-moisture content, infiltration capacities, permeability of aquifers, height of the phreatic surface, etc.

Catchment characteristics form the stage on which the drama of water-balance and run-off are performed. The exact values assumed by the water-balance or the run-off at any given time are externally controlled in that they depend on both the energy and the precipitation falling upon the catchment. In particular, run-off is a direct response to precipitation.

From the discussion so far it should be clear that the hydrograph can be separated into at least two components, which are the base flow and the direct run-off. Under dry weather conditions the flow will consist completely of base flow while under heavy rain it will be dominated by the direct run-off. The separation of base flow from the direct run-off in hydrographs is difficult and usually rather arbitrary. A good method is to project the base flow prior to the storm below the hydrograph allowing for a small increase in the base flow during the storm.

Probably the most important contribution made to rainfall/run-off studies was the concept of the 'unit hydrograph' which was proposed by L.K. Sherman in 1932. Sherman sought a correlation between net or effective rainfall, that is, the water remaining as run-off after all losses by evaporation, interception and infiltration, and the surface run-off, that is the total

TABLE 8.1
DERIVATION AND APPLICATION OF 12 HR S-CURVE HYDROGRAPH

Time (hr)	12-hr Unit Hydrograph	12 hr S-Curve	S-Curve Shifted 24 hr.	Col. 3 − Col. 4	24-hr Unit Hydrograph (Col. 5/2)
(1)	(2)	(3)	(4)	(5)	(6)
0	0	0		0	0·0
6	7	7		7	3·5
12	11	11		11	5·5
18	13	20		20	10·0
24	15	26	0	26	13·0
30	12	32	7	25	12·5
36	8	34	11	23	11·5
42	6	38	20	18	9·0
48	4	38	26	12	6·0
54	0	38	32	6	3·0
60		38	34	4	2·0
66		38	38	0	0·0
72		38	38	0	0·0

(a) All flows in m^3/sec.

(b) Column 2 represents the run-off from 25 mm net rainfall over 12 hours.

(c) Column 3 is obtained by lagging the unit hydrograph in column 2 by successive 12-hr periods and adding up the values for each observation time. It represents the run-off from a continuous rainfall at the net rate of 25 mm per 12 hr (i.e. 50 mm per 24 hr).

(d) Column 5 represents the difference between columns 3 and 4 and is the run-off from 50 mm over 24 hr (i.e. 25 mm over two periods of 12 hr).

(e) Column 6 is the 24-hr unit hydrograph and is obtained by dividing column 5 by 2 since it represents the run-off from a rainfall of 25 mm over 24 hr.

hydrograph minus base flow. The unit hydrograph is the hydrograph of surface run-off from a given catchment due to a specified uniform net rainfall over a stated unit of time. It will, therefore, represent the integrated effects of all the catchment constants such as drainage area, channel slope, pattern of stream channels, channel capacities, land slopes and other physical factors, since for a given net precipitation the storm characteristics have been eliminated, and for a given catchment and net rainfall it should be constant in shape. The net rainfall is almost invariably taken as 25 mm (1 inch) over the contributing catchment, while the unit of time for the duration of rainfall depends upon the area of the catchment but it should not be less than the time of concentration. The unit of time may be a day or more for larger catchments or 6 or 12 hours as the size of the drainage area decreases. If the duration of rainfall is selected as 6 hours, then the resulting unit hydrograph is referred to as the 6-hour unit hydrograph, and similarly for rainfalls of other durations.

If it is assumed that the excess rainfall continues indefinitely, with uniform intensity and areal distribution, then the hydrograph will reach a constant value, and the curve obtained is known as an S-curve hydrograph. The S-curve is a useful theoretical tool, allowing various unit hydrographs to be calculated. Thus if two identical S-curves are offset by a given time interval and subtracted, then the resulting hydrograph is the unit hydrograph for the rainfall with a duration equal to the offset time interval. Hydrographs for measured net rainfall amounts other than 25 mm may be obtained by multiplying the flow values of the appropriate unit hydrograph by the ratio of the measured net rainfall to 25 mm. The uses of S-curves are illustrated in Table 8.1.

Some Further River Hydraulics

Natural river channels are in general very irregular, usually varying in section from an approximate parabola to an approximate trapezoid, and in streams subject to frequent floods, the channel may consist of a main channel section carrying normal discharges and one or more side channels for accommodating overflows. Some definitions of important geometric elements of the river channel (Figure 8.3) are discussed below.

(i) The depth of flow is the vertical distance of

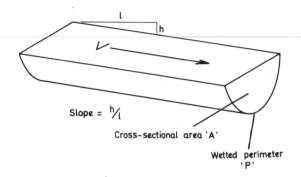

Figure 8.3 Stream geometry

the lowest point of a channel section from the free-water surface.

(ii) The stage is the elevation or vertical distance of the free-water surface above a given datum. If the lowest point in the channel section is chosen as the datum, the stage is identical with the depth of flow.

(iii) The water area A is the cross-sectional area of the flow normal to the direction of flow.

(iv) The wetted perimeter P is the length of the line of intersection of the channel-wetted surface with a cross-sectional plane normal to the direction of flow.

(v) The hydraulic radius R is the ratio of the water area to its wetted perimeter,

$$R = \frac{A}{P}$$

Water velocities in a river are not uniformly distributed in the channel section owing to the presence of a free-water surface and to the friction along the channel wall, and the maximum velocity often appears at some distance below the surface. The velocity distribution also depends on the shape of the section, the roughness of the channel, and the presence of bends. The measurement of water velocity in streams and rivers is normally accomplished by means of a current meter, of which there are two main types. In the first, the flowing water acts upon a propeller which is attached to a horizontal spindle, while in the second it acts upon an arrangement similar to a cup anomometer, that is a number of cups or buckets attached to a vertical spindle. The rotation of the spindles causes the closing of a magnetic contact which results in a small current flowing in an electric circuit and thus advancing a digital counter. There usually exists for the instrument a calibration curve which relates the number of revolutions per second to the velocity of the water. A current meter can only give an accurate indication of the water velocity at a single point in the stream, and since it has already been stated that water velocity is highly variable within a channel section, this single reading will not be typical of the mean velocity within the stream. A number of readings at various depths across the stream are required for an estimate of the mean velocity. If the shape of the channel section is known, together with the stage, it is possible to calculate the cross-sectional area of the stream, and then if the mean velocity is measured, to compute the river-flow.

It would be useful if at a known cross-section of the river channel it were not necessary to measure the mean water velocity before computing the river-flow, that is if river-flow could be estimated from a knowledge of the river stage alone. The mean river velocity may be estimated from the Manning formula, which is:

$$V = \frac{R^{\frac{2}{3}} S^{\frac{1}{2}}}{n}$$

where V is the mean velocity in m sec^{-1};

R is the hydraulic radius in m;

S is the slope of the water surface;

and n is the coefficient of roughness, often known as Manning's n.

The value of n in most streams decreases with increasing stage and discharge. When the water is shallow, the irregularities of the channel bottom are exposed and their effects become pronounced. At low flows n typically has a value of between about 0·05 to 0·150 depending on the unevenness of the river-bed. As the stage increases, n falls in value reaching a minimum of about 0·03 when the stream is just about to overflow its banks. When the river is in flood it may overflow its banks and a portion of the flow will be along the flood plain which generally has a higher n value than that of the channel proper. Thus once the river overflows its banks the mean value of n starts to increase. In artificial channels where the bed and banks of the channel are equally smooth and regular, the value of n may remain nearly constant for all water-levels. The slope and mean section of a reasonably uniform stretch of river may be found by a simple topographical survey, and it is then possible to estimate the velocity for every stage from the

Manning formula and therefore the corresponding river-flow. Once a stage-discharge relationship has been obtained for a given river section, a continuous record of the stage by means of a water-level recorder will allow the flow at any time to be estimated.

A water-level recorder at a carefully surveyed segment of river constitutes a very simple river-gauging station. For more accurate and stable results it is better to replace the natural river channel by an artificial channel with known flow characteristics. Such an artificial channel is known as a weir or flume. The high flows in a river may be more than a thousand times greater than the low flows, and so the shapes of weirs have to be complex if they are to maintain a stable stage-discharge relationship over a wide range. Furthermore, they have to withstand damage when the river is in flood, be unaffected by the accumulation of silt, and allow fish to pass. Since at the weir there is a precise and stable relationship between stage and discharge, a water-level recorder placed just upstream of the weir will give a continuous record of the river-flow.

Floods

A flood may be defined as an unusually high rate of surface run-off, often leading to the inundation of large areas of land particularly near rivers. Natural floods are nearly always the result of surface, rather than groundwater, run-off, and are thus caused by intense rainfall, sudden snowmelt, or a combination of these two events. Floods can also be caused by non-meteorological events, such as the sudden collapse of dams, but the latter are outside the context of the discussion which is restricted to natural floods.

The analysis of rare events, such as floods, may be approached by compiling the data for analysis in two ways. The most commonly used series in extreme value analysis of this type is the annual series, which consists of the maximum flood event for each year. In contrast, a partial duration series is one in which are listed only events with magnitudes above a selected base which could be the mean annual flood. A limitation of the annual series is that each year is represented by only one event, but the second highest event in one particular year may be higher than the highest recorded in some other year, yet not be counted. Unfortunately, the events in the partial duration series may not be truly independent, and this can lead to difficulties in the statistical analysis of the series.

The frequency of occurrence of floods is often considered in terms of the return period. If p is the probability of the occurrence of a value for an annual maximum flood event equal to or greater than a given value x, then the return period T is defined by:

$$T = \frac{1}{p}$$

For the annual series, the return period is the average number of years which must elapse before the magnitude of the flood will be equalled or exceeded. The return period for a partial duration series is the average interval between floods of a given size regardless of their relation to the year as a period of time. As a result, the return period of an event with a given magnitude in a partial duration series will be shorter than for the same event in an annual series, though in practice, for return periods exceeding 10 years, the distinction between annual and partial series is inconsequential.

The probability (P_T) that a flood will occur within its own return period is given by:

$$P_T = 1 - \left(1 - \frac{1}{T}\right)^T.$$

The solution of this equation for typical values suggests that the probability that a flood will occur within its own return period is about 64 per cent. This is an interesting result, because dams are often designed only to withstand floods with return periods less than some specified return period, and for a dam designed to withstand a flood with a 25-year return period there is a 64 per cent chance that the flood will be exceeded before the end of the first 25-year period. For design purposes, it is better to specify some acceptable risk that a flood will occur within the designed life of the structure and then to calculate the required return period of the flood. For example, if it is desired to construct a dam with a 90 per cent confidence (i.e. 10 per cent risk) that its flood capacity will not be exceeded within the next 25 years, then it will be necessary to design for a flood with a 240-year return period. Most dams, river bridges, etc., are built to last for at least 100 years and if a permissible risk of failure of 1 per cent is allowed, this means that they must withstand a flood with a 10 000-year return period.

Magnitudes of floods with relatively short return periods may be estimated by statistical procedures, but estimates become uncertain if the return period exceeds about four times the length of the river-flow record. Further difficulties may arise if no river-flow records exist for a particular catchment, and yet flood estimates are required. Rainfall records are normally far more common than river-flow records and it has already been demonstrated by the discussion of hydrographs that river-flow may be predicted from rainfall. Flood estimates may therefore be improved by a statistical analysis of rainfall records. Statistical procedures are unsuitable for estimating flood or rainfall magnitudes with very long return periods such as the 10 000-year period mentioned earlier, and in this case the concept of the probable maximum flood and the probable maximum precipitation have to be used. There is no universal agreement on a simple precise meaning of probable maximum precipitation, but the definition that is often used is that it is the rainfall depth, for a particular size of catchment, that approaches the upper limit that the present climate can produce. Some hydrologists, however, think of it only in terms of flood flows and they would define it as that magnitude of rainfall over a particular catchment which will yield the flood flow of which there is virtually no risk of being exceeded. The latter is, of course, the probable maximum flood. These definitions imply that there is a physical upper limit to the amount of rainfall in a given climatic zone. This is so because physical restrictions on the joint occurrence of the various rain-forming meteorological mechanisms impose an upper limit on the rainfall magnitude.

A pioneer and major worker on the statistical analysis of extreme values was Professor E.J. Gumbel. He showed that the magnitude (y) of an intense rainfall or extreme flood which may be expected to be equalled or exceeded on the average only once every T years is given by:

$$y = \overline{y} + k\sigma;$$

where $\overline{y}$ and σ are the mean and standard deviations of the observed annual maximum rainfalls, and k is a constant which depends on both the sample size and the return period T and may be found from tables.

Return periods of individual values in a series of annual maxima may be estimated in the following manner. The data, perhaps the largest daily rainfalls in each year, are first arranged in descending order of magnitude and assigned an order number m, where $m = 1$ for the largest value and $m = n$ (the number of items in the sample) for the smallest value. The return period T of an individual value is then given by:

$$T = \frac{(n+1)}{m};$$

where the terms are as previously defined. Results may be plotted on semi-logarithmic graph paper, a logarithmic scale being used for the return period (T) and a normal linear scale for the magnitude (y). The values (Figure 8.4) are found to lie along a straight line which is given by Gumbel's equation. It is therefore possible by the use of the above two equations to determine, within the limits of the record, the magnitude of a flood or rainfall with any given return period.

Probable maximum precipitation magnitudes may be estimated by the technique of maximizing the storm rainfalls which occur over the catchment. The starting-point is the identification of storms which have occurred over the study catchment or over catchments which are geographically and climatically

similar. These storms are optimized to produce the maximum rainfall by a careful adjustment of the rain-forming mechanisms within the limits of climatic possibility. The most important rainfall-limiting mechanism is the flow of moisture into the storm, and this usually sets an upper limit to the storm precipitation.

For a flood to approach the probable maximum value, a number of conditions must be satisfied.

(i) The duration of the storm must exceed the time of concentration of the catchment.

(ii) The storm must cover the whole of the catchment.

(iii) Over the time specified in (i) and the area specified in (ii), the storm must approach the probable maximum intensity.

Unless the storm duration exceeds the time of concentration of the catchment the hydrograph of river-flow will not reach a stable maximum value, but will be truncated before it reaches its full potential as is normally observed. Similarly, the storm must cover the whole catchment so as to produce the maximum run-off. Also the intensity of rainfall tends to decrease as the area covered by the storm increases. Storm size and duration are therefore important when considering the possibility of a flood. A small storm

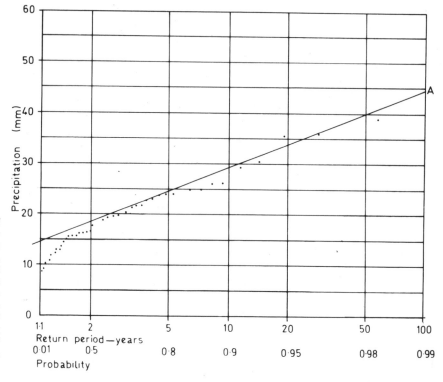

Figure 8.4   One-day rainfall magnitude-frequency relationships. The vertical linear scale refers to rainfall magnitudes, while both probabilities and equivalent return-periods are shown on the horizontal logarithmic scale. Dots refer to individual values in a series of annual maxima. The straight line A is obtained by use of Gumbel's equation which was described in the text.

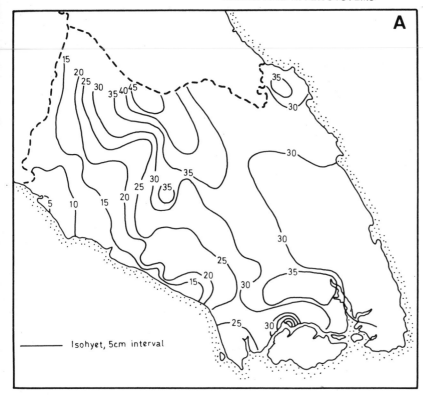

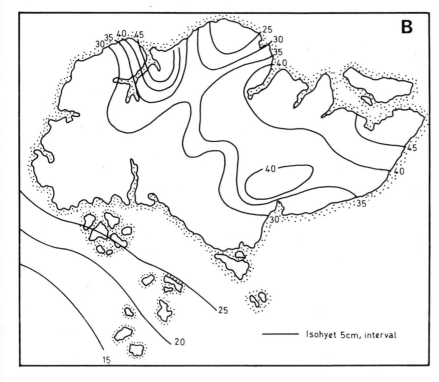

Figure 8.5 Two-day total rainfall distribution, 9th and 10th December 1969.

A. Johore

B. Singapore

(After Chia Lin Sien and Chang Kin Koon (1971), 'The record floods of 10th December, 1969 in Singapore', *Journal of Tropical Geography*, 33, pp. 9).

may produce a relatively large flood in a small catchment but produce little change in the run-off from a larger area. In particular, thunderstorms produce deep floods over small areas but have little effect on the flow of large rivers which are only affected by widespread rains. This is often observed in tropical countries. Thunderstorms are frequent in Malaysia, causing local flooding of both flat land and small catchments. In the case of small steep catchments the flooding can cause both severe erosion and extensive local damage, but these are always very restricted in area. Widespread extensive rains sometimes occur along the north-east coast of Peninsular Malaysia between November and March, giving rise to extensive floods on rivers flowing into the South China Sea. A good example of this occurred in early January 1967, when floods of great magnitude occurred in the Malaysian States of Kelantan, Trengganu and Perak. A large storm developed in the South China Sea and moved into Kelantan and Trengganu giving intense and widespread rain in these two states, some places recording falls of 60 cm (24 inches) in 24 hours. The extensive rains caused floods in the coastal regions of the two states, the inland areas being spared. On 4 January, a storm centre moved across the 'saddle' on the Thailand border into the upper catchment of the Perak River and caused flooding in the river which flows towards the west coast. These major river floods were caused by rainfall intensities which even at their maximum value did not approach those in an average thunderstorm (7·5—10 cm per hour), thus illustrating the importance of the long duration and widespread extent of the rain.

On a smaller scale, heavy rainfalls during 9 and 10 December 1969, caused damage to a total of about M$ 4·5 million in Singapore. The rainfall distribution on this occasion is shown in Figure 8.5, where the return periods vary from 20 years in the north of Singapore Island to 5 000 years in the east.

With high rainfall totals, the condition of the soil is unimportant, since even if it were completely dry at the start of the storm, it would soon be saturated and flood-producing run-off would result. If the rainfall is marginal for flood formation, the following conditions may be helpful to the production of a flood.

(i) The soils in the catchment should be nearly saturated with water and the infiltration capacity of the catchment should be low.

(ii) The interception capacity of the vegetation within the catchment should be low.

(iii) There should already be a significant flow in the river.

Moderate rain on a catchment with these particular characteristics will probably produce a flood, while a similar rainfall on a very dry catchment will produce no significant run-off.

Transportation, Deposition and Erosion by Rivers

The denudation of the continents depends largely on the ability of running water to carry solid matter and minerals in solution. A solid object at the surface of the earth experiences a downward force equal to its mass times the acceleration of gravity, but if the object is placed in water, this downward force is reduced by the buoyancy of the water. According to Archimedes' principle, the downward force is reduced by the force acting on a volume of water equal to that of the object. As the object moves through the water, the water must flow to permit it to sink, and exerts an upward force known as the drag of the water on the object. When an object starts to sink, it accelerates until the force of the drag is equal to the downward force, and at this point the rate of fall becomes constant. The value of the constant sinking rate varies with the size of the particle, thus for sand grains of diameter 1 mm, it is about 1 mm per s, while for silt particles of diameter 0·01 mm it is only 0·1 mm per s. These rates apply to still water, but in a flowing river there are numerous turbulent eddies and these may make it almost impossible for fine particles to settle out.

Moving water is able to transport particles in two ways; firstly, it carries fine silt and clay particles in suspension and, secondly, larger particles are dragged along the bed of the river. The particles carried in suspension often give river water a brown colour but they slowly settle out if the water is allowed to stand. Turbulent motions above the stream bed impart a drag force on grains resting on the bed and when the drag becomes sufficiently great, the grains start moving. So the faster the river flows the greater is its potential for transporting solid material. Rivers will therefore erode when the flow is fast, that is where the river-bed is steep or when the stream is in flood, and deposit material when the flow is slow, that is the slope is gentle or the stage is low.

Water is also an excellent solvent and rain-water falling on the land partially dissolves rocks and carries the dissolved salts to the sea via river-flow. Some substances are extremely soluble in water, salt (NaCl) being a good example, but most solids are at least slightly soluble, so river water will contain a great variety of dissolved minerals.

Observations of streams suggest that the amount of material carried both as dissolved solids and in suspension depends very largely on the climate and on the nature of the catchment. Erosion rates are often high in catchments which are being greatly disturbed by building or open-cast mining but this discussion is restricted to relatively undisturbed natural catchments. Chemical weathering and solution take place most readily at high temperatures and it is therefore a common assertion that the humid tropical environment favours chemical weathering, but this is not necessarily so since Douglas (1967) has found that streams draining tropical forest-covered granite catchments in Singapore, Peninsular Malaysia and north-east Queensland carry more of their total load as suspended matter than matter in solution.

Douglas considers that the most important factor affecting the total sediment load of streams draining rain forest-covered catchments is the frequency of intense storms, because streams erode most when in flood. Intense storms giving rise to floods are infrequent, so rivers have very low flows for long periods of time and in this state erosion is slight. The rare floods caused by intense storms cause most of the erosion within catchments and therefore give rise to the highest sediment loads. Douglas has illustrated this for the Behana in north-east Queensland, where 50 per cent of the total sediment load is carried during seven days of the year. Actual total sediment loads for any given flow are highly variable and depend on the immediate hydrological history of the catchment. Thus seasonal drying of soils causes decreasing stability, and intense erosion results from the first heavy rains of the wet season, but once the loose material has been removed, later storms may cause considerably less erosion.

The long profiles of many rivers are concave upwards, that is with the steepest slopes near the source. Davis has argued that eventually streams attain equilibrium longitudinal profiles in which, at any

point, the energy of the stream is just sufficient to transport all the debris arriving from upslope, but insufficient to engage in further downcutting at that point, provided that average conditions are considered. The uncertain and variable nature of river-flow and sediment transport suggests that although this explanation of the typical longitudinal profile of streams is generally true, it may be over-simplified.

FURTHER READING FOR CHAPTERS 7 AND 8

Begemann, F. and Libby, W.F. (1975). 'Continental water-balance, ground water inventory and storage times, surface mixing rates and world-wide water circulation patterns from cosmic-ray and bomb tritium.' *Geochemica et Cosmochimica Acta., 12,* p. 277.

Bruce, J.P. and Clark, R.H. (1966). *Introduction to Hydrometeorology* (Pergamon Press, Oxford).

Davis, W.M. (1954). *Geographical Essays* (reprint) (Dover Publications, New York).

Douglas, I. (1967). 'Erosion of granite terrains under tropical rain forest in Australia, Malaysia and Singapore.' Symposium on River Morphology, International Union of Geodesy and Geophysics, General Assembly of Bern, Sept.-Oct., 1967. p. 31.

Eyles, R.J. (1967). 'Laterite at Kerdan, Pahang, Malaya.' *Journal of Tropical Geography*, 25, p. 18.

Horton, R.E. (1939). 'Analyses of run-off plot experiments with varying infiltration capacity.' Transactions of the American Geophysical Union, Part IV, p. 693.

Kenworthy, J.B. (1970). 'Water and Nutrient Cycling in a Tropical Rain-forest.' In Flenley, J.R. (ed.), *The Water Relations of Malesian Forests*, University of Hull, Department of Geography, Miscellaneous Series, II.

Niewwolt, S. (1965). 'Evaporation and Water Balances in Malaya.' *Journal of Tropical Geography*, 20, p. 34.

Penman, H.C. (1948). 'Natural evaporation from open water, bare soil and grass.' Proceedings of the Royal Society, London. Ser. A. 1933, p. 120.

Thornthwaite, C.W. (1948). 'An approach toward a rational classification of climate.' *Geographical Review*, 38, p. 55.

Wiesner, C.J. (1970). *Hydrometeorology* (Chapman and Hall, London).

9 Glaciers and Glaciation

EXTENSIVE snow-fields are common in winter over the interiors of middle-latitude continents, since much of the precipitation falls as snow and because of the low temperatures, the snow remains on the ground until it is melted by rising temperatures during the following spring. If all the snow did not melt during the following summer a permanent snow-field would form, and the snow would slowly compact to form an ice-sheet. Since permanent snow and ice are only found on high tropical mountains, it could be assumed that they are of only limited interest in any discussion of the tropics. Actually, they are of great importance because recently (within 15 000 years) the middle latitude ice-sheets were much more extensive than at present and caused both climatic changes and a lowering of sea-level by up to 150 m within the tropics. Thus the middle latitude glaciations of the last million years have had a large influence on tropical landforms and ecosystems. Within large areas of the temperate land masses one result of the recent glaciations was to denude the landscape of both soil and vegetation; so many temperate ecosystems are only about 10 000 years old and are considerably younger than their tropical counterparts. Therefore part of the difference between the soils in particular and the vegetation communities of the temperate latitudes and the tropics results from the considerably greater age of the tropical formations.

Snow- and ice-fields tend to generate their own local climates because snow and ice have a high albedo for short-wave radiation but act as excellent blackbodies in the far infra-red. This means that snow reflects most of the sunlight falling upon it and also that it cools very effectively by radiating in the far infra-red. As the sun is not able to warm the snow surface while the snow itself is slowly cooling, temperatures are lower than they would be if the ice-sheet were absent. The low surface temperatures over an ice-sheet often lead to the formation at the surface of a shallow layer of cold air which is restricted in depth and is usually overlain by warmer air in a temperature inversion. Because of the intense cold the surface air is slightly denser than normal, and this can give rise to the impression that a shallow anticyclone overlies the ice-sheet, but the normal atmospheric circulations take place above the cold surface layer.

Intense cold alone is not enough to form an ice-sheet, for in addition there must be a large excess of winter snowfall over summer melting. This implies a copious supply of moisture for snowfall which can only come from a nearby warm ocean. The Greenland ice-sheet is a good example since it is surrounded by the relatively warm North Atlantic Ocean.

In general three different major types of glacial climate may be distinguished:

(i) A widely distributed high-polar ice-cap climate having summers below 0°C and winter temperatures of −20°C to −40°C e.g. North Greenland and Antarctica.

(ii) A continental type of tundra climate with cool summers (warmest month below 10°C), cold winters (coldest month below −8°C), wide temperature fluctuations, small precipitation, saline efflorescence, and wind action and rock weathering similar to that of arid regions e.g. West Greenland.

(iii) A maritime type with small oscillations of temperature and cool summers, e.g. South Georgia, South Orkneys, Spitsbergen and Iceland.

To these types should be added the climates of high tropical mountain peaks which can be snow-covered and may support glaciers.

Snow, ice and glaciers can have a profound effect on the landscape, since they produce distinctive erosional features and also distinctive deposits which can be studied in existing glacial areas. More interesting is the fact that similar features are found in areas far distant from existing glaciers and enjoying mild climates, and this can be taken as evidence that these areas were covered by ice-sheets and glaciers in the not very distant past. In the geological record, evidence is found of the existence of glacial periods in the far geological past, for example, evidence of a major glaciation dating from the Carboniferous is found in India. It will be shown later that the last glacial period indirectly affected even the tropics including South-East Asia.

The Elementary Physics of Ice and Snow

If the surface temperature is near or below $0^{\circ}C$, precipitation will take the form of snow which will settle on the ground, and if the temperature is low enough the snow cover will remain throughout the year forming a perennial snow-field. Perennial snows on mountains and in polar regions are bounded by a lower line, the so-called snow-line, where as much snow melts in summer as falls in winter. Since the

level of the snow-line is controlled by the snowfall and by the energy available to melt or evaporate the snow, it will not therefore correspond to any particular temperature. In practice it appears to oscillate between mean annual temperatures of $10^{\circ}C$ and $-10^{\circ}C$; on Kilimanjaro it lies at $-4^{\circ}C$ and in central Asia between $-6^{\circ}C$ and $-8^{\circ}C$.

Basically the perennial snow-line represents a crude balance between accumulation and ablation. Accumulation may be defined as the substance added to an ice or snow surface by snowfall, hail, frost, rain that freezes, and drifted and avalanched snow. Ablation is the substance lost from an ice or snow surface by melting, evaporation and sublimation, deflation and calving.

Ice-sheets form where snowfall exceeds annual ablation by the compaction and structural alteration of the snow to ice. The density of fresh snow is about $0 \cdot 15$ to $0 \cdot 16$ g cm^{-3}, but after settling, removal of part of the air and after recrystallization, the stage of granular snow or firn (density about $0 \cdot 5 - 0 \cdot 8$ g cm^{-3}) is attained. Repeated melting and refreezing, aided by further compaction under pressure from overlying

Figure 9.1 Structure of a valley glacier. The bergschrund is a characteristic crevasse at the head of a valley glacier. The firn limit is the lowest altitude of annual net accumulation and is the equivalent of the snowline on nearby mountain sides. (After Bloom, 1969).

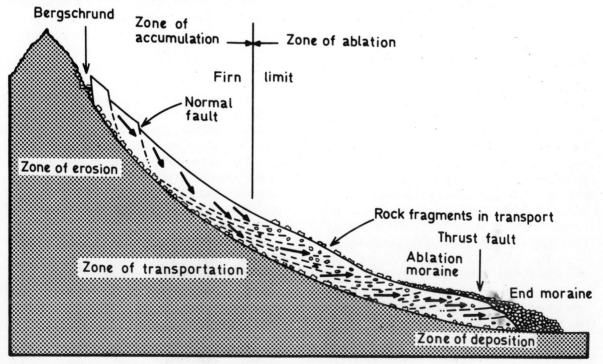

Bergschrund

Zone of accumulation — Zone of ablation

Firn limit

Normal fault

Zone of erosion

Rock fragments in transport

Thrust fault

Ablation moraine

End moraine

Zone of transportation

Zone of deposition

firn leads to complete impermeability to air and densities exceeding 0·82 g cm^{-3}. This resulting product is defined as ice and the resulting ice masses may form either mountain, valley or piedmont glaciers in areas of rough highland terrain, or ice-caps in areas of smoother topography.

If accumulation exceeds ablation, there must be a net gain in the amount of surface ice. This positive balance of accumulation over ablation is removed by the flow of ice into areas where ablation exceeds accumulation; such a flow of ice forms a glacier (Figure 9.1), and since the glacial ice is in motion it can descend below the perennial snow-line. A glacier will terminate at the point where ablation equals accumulation plus ice flow into the area. Clearly, glaciers will not necessarily terminate at a level corresponding to some particular temperature because their general level of activity will be controlled by the integrated results of snowfall and the factors influencing ablation, such as air temperature and solar radiation. An advance in a glacier can be brought about by a marked increase in snowfall as well as by a fall in temperature. Similarly, glacial retreat can be caused by falling snowfall amounts even if the temperatures remain constant.

Velocities at any point on a glacier are subject to marked day-to-day variations, reflecting short time period variations in accumulation and ablation, but it is nevertheless possible to obtain average values of glacier motions that are meaningful in that they indicate the scale of the phenomenon. Annual velocities in the bigger Alpine glaciers vary between 30 m and 150 m. The Hoffellsjokull Glacier in Iceland appears to move at an average annual rate of 225 m, and may have a maximum annual speed of 700 m. The average daily velocity of some Himalayan glaciers may be about 50 m. Antarctic ice, on the whole, moves very slowly; the Beardmore Glacier, in its swiftest part, flows at about 9 m per year, the Ferrar Glacier at about 0·45 m, the Blue Glacier, Royal Society Range at about 1·2 m. Considerations of precipitation, ablation and flow suggest that the snow in the interior of the Antarctic may reach the margin only after 20 000, 25 000 or even 50 000 years.

The temperature structure within a glacier may have an influence upon glacial activity. Glaciers show two temperature zones, an outer one of fluctuating temperatures and an inner one in which the temperature is approximately uniform. In the outer zone, the ice behaves like any other rock and its temperature is susceptible to oscillations governed by changes in air temperature and solar radiation; each day sends a wave of warmth down into the glacier, each night a cold wave. The depth to which these oscillations penetrate varies with locality, e.g. in Greenland they have been measured to a depth of 50 m. A deeper zone of constant temperature is present, unless the ice is thinner than the zone of fluctuating temperature. The upper part of the zone of constant temperature has a temperature approximately equal to the mean annual temperature of the air. Where this is not far from 0°C, the lower layers of the glacier will be approximately at freezing-point, e.g. in the Alps. If, as in high latitudes, the annual temperature is far below 0°C, the glaciers are said to be 'cold'; thus in Greenland the zone of constant temperature ranges between −15°C and −32°C, and similar observations have been made in Antarctica.

Temperatures at great depths in glaciers tend to increase by about 1°C for every 20 m until they approach the melting-point, because of the heat generated by internal friction due to ice movement and also to the supply of heat from the earth beneath the ice. The bottom layers of glaciers in contact with the solid rocks are probably roughly at melting-point, but exceptions occur where the ice is thin and in extremely high-polar glaciers. In the average glacier, except in the high altitude parts, the ice rests on ground that is always above freezing-point.

Glaciers can be classified geographically into 'temperate' or 'warm' glaciers and 'polar' glaciers, and the latter can be further subdivided into high-polar and sub-polar glaciers. In temperate glaciers the temperature is near melting-point even in winter, and the top few metres only have a significant negative temperature. Even in summer polar glaciers have a marked negative temperature to a depth of about 100 m; the high-polar type has no melt-water at the base, and the sub-polar type only melts at the base during summer.

The causes of ice movement are complex. The actual velocity of movement will depend on a variety of factors including accumulation and ablation, the gradient, the volume and depth of the ice, basal and marginal friction, and the internal temperature and general cohesion of the ice. In many ways ice is a paradox for it possesses a clearly crystalline structure, and yet it appears to move as a plastic substance in that it can deform and flow.

Glacial Erosion and Deposition

Although the action of glaciers can be studied in existing glacial regions, the Pleistocene ice-sheets have left traces over vast areas of the middle latitudes and on high tropical mountains that can also be examined. Since glaciers are powerful agents of erosion and deposition their former presence is shown by their deposits and by the typical landforms they produce. Glacial deposits often provide this type of evidence because they last far longer than purely erosional features which can be quickly modified or removed. Glacial landforms include striations, roches moutonnées, drumlins, U-shaped valleys, cirques, terminal moraines, kames, kettles, eskers and valves.

STRIATIONS

Glaciers can act as agents of erosion by the general abrasive action of ice and ice rock mixtures on the bed rocks. The most common and conspicuous unit of glacial abrasion is the striation (Stria or scratch). Striations are fine cut lines on the surface of the bed rock which are inscribed by rock fragments in the overriding ice. They can vary in size from grooves 1–2 m in depth and 50–100 m long to very fine striations that grade downwards into a general polish, but not all striations on rock are of glacial origin.

ROCHES MOUTONNEES

Sometimes hillocks in glaciated regions are assymmetrical, the gentle side facing upstream and the steep side facing downstream. Whereas the upstream sides are commonly abraded, the cliff-like downstream sides generally coincide with joints or other minor structures and are rarely abraded. Evidently the cliffs are the work of mechanical erosion, termed glacial quarrying (or glacial plucking) by analogy with the lifting out of blocks in a quarry. These asymmetrical features are sometimes termed 'roches moutonnees', and their asymmetry affords clear evidence of the direction of glacier movement.

DRUMLINS

Alternating glacial erosion and deposition can produce a whole family of streamline features of which the best known is the drumlin, an Irish word (Gaelic druim, the ridge of a hill) first used as a general term for hills of this shape by Maxwell Close in 1866. The 'ideal' drumlin form is half-ellipsoidal like the inverted bowl of a spoon, with the long axis paralleling the direction of flow of the former glacier. Most drumlins vary from the ideal, grading from rounded hills into long narrow shapes with sharply pointed upstream ends, or into bed rock knobs with accumulations of glacial deposits streaming from their lee sides—a combination known as crag and tail. Though many drumlins consist of glacial deposits only, some have cores of bed rock, usually at or near their upstream ends, while others consist almost entirely of bed rock, thus creating a complete graduation, independent of outward form, from solid rock to glacial deposits.

U-SHAPED VALLEYS

The features studied so far are rather minor in form, but sometimes very large landscape features result from glacial sculpturing. Although valleys modified by glaciation occur in many parts of the world, probably none of them are primarily the work of glaciers but rather are pre-existing valleys remodelled by glaciers. Glaciers do not cut valleys in areas of low relief but instead they spread out and erode the surface with broad uniformity. The simplest way to analyse the erosive effects of glaciers on valleys is to compare glaciated valleys with normal stream valleys. In general the glaciated valley has a step-like long profile, steeper in the headward part than that of a stream valley and between the steps there may be rock basins, which can reach large depths near the lower ends of the glaciated valley. If the glaciated valley is subsequently flooded by a rise of sea-level a deep inlet known as a fiord will be formed.

In most glaciated valleys a large amount of material is removed by widening, and this results in their having a pronounced U-shaped cross-profile. Some stream valleys have a U shape, but it develops slowly and the side slopes are graded, whereas the glaciated U valley develops rapidly and the side slopes are not graded. Most glaciated U valleys are made by alteration of youthful V valleys originally cut by streams and mass-wasting. Glacial erosion tends to remodel a valley towards a semi-circular cross-section because this shape offers least frictional resistance to the flowing ice. The deepening or widening of the main valley by glacial erosion may proceed at a faster rate than that in tributary valleys, thus leaving the tributary valley floor, at its junction with the main valley, higher than the main valley floor and upon deglaciation, the stream in the tributary falls or cascades down to the main valley floor. Such tributaries which are called hanging tributaries, are normal in glaciated valleys because of differences in ice thickness.

CIRQUES

The heads of many glaciated valleys are shaped like a theatre or half a bowl, and this feature is widely known as a cirque (or corrie). It occurs not only at the head of a valley but also independently as a major indentation in an otherwise smooth slope. A cirque can be described as a deep, steep-sided recess, roughly semi-circular in plan, cut into a slope by erosion beneath and around a bank of firn or a glacier. It may reach a diameter of much more than 1 km, and its floor may contain a rock basin. Since many cirques are deeper near their headwalls than in their outer parts, the result is that their floors form rock basins that may contain lakes. At least two groups of processes are involved in the formation of cirques; firstly, frost wedging accompanied by mass-wasting, and secondly, glacial erosion, but the first process is normally the most important. The essential features of the process are melting by day and freezing by night during the summer ablation season beneath a small snow patch; the fragments shattered by frost action move downslope by creep. This combination of thaw-and-freeze with mass-wasting is called niva-

tion. Cirques form at or close to the orographic snow-line, which normally lies at or slightly above the cirque floor.

A highland affected by a glacial climate is sculptured in a characteristic manner known as Alpine sculpture, of which the main features are shown in Figure 9.2. Under normal non-glacial erosion, mountains have smooth contours and valleys are broadly V-shaped. The growth of cirques in a glaciated mountainous region produces a dramatic landscape, with sharp ridges and pyramid peaks, and their continued growth on opposite sides of a crest eventually reduces the crest to a knife-edge form (an arête). Three or more cirques gnawing inward against a high mountain crest produces a pyramid (a horn).

In Europe, North America and Asia large areas are covered by so-called glacial drift—a term which embraces all rock material in transport by glacier ice, all deposits made by glacier ice and all deposits predominantly of glacial origin made in the sea or in bodies of glacial melt-water, whether rafted in icebergs or transported in the water itself. Clearly there can be a very large range of drift features, and a few of the more important ones will now be considered.

MORAINES

A moraine is an accumulation of drift whose topography is independent of the surface underneath it,

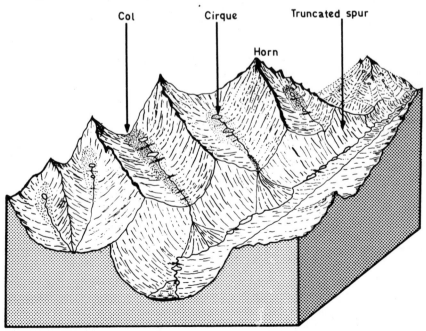

Figure 9.2 Alpine landscape. Various landscape features produced by glacial erosion in a mountainous area. (After Bloom, 1969).

and which has been built by the direct action of glacier ice. There are two major kinds of moraine, ground moraine and end moraine. Ground moraine which has low relief devoid of major ridges, forms undulating plains possessing a local relief of less than about 6 m, and normally has accumulated beneath a glacier. If an end moraine, which is a ridge-like accumulation of drift built along any part of the margin of a glacier, is built along the downstream margin, it is called a terminal moraine, and if it is built along the lateral margin of a valley glacier, it is known as a lateral moraine.

Ice contact stratified drift can form upon, against, or underneath the wasting terminal zone of a glacier. The same site may see a large variety of conditions ranging from a rushing stream to a pool or overriding ice. Clearly ice contact stratified drift could take almost any form and some of its more common forms include kame terraces, kames, kettles and eskers.

KAMES

A kame terrace is an accumulation of stratified drift laid down chiefly by streams between a glacier and an adjacent valley wall and left as a terrace after the disappearance of the glacier. The significant distinction between kame terraces and ordinary stream terraces is that the former are not the remnants of former flood plains that covered the whole valley floor, for they were never much more extensive than they are now.

Kames are mound-like hills of ice contact stratified drift, some of which are bodies of sediment deposited in crevasses and other openings in or on the surface of nearly stagnant ice which later melted leaving the accumulated sediment in the form of isolated or semi-isolated mounds.

KETTLES

A kettle is a basin in drift, created by the ablation of a former mass of glacier ice that was wholly or partly buried in the drift. Most kettles are less than 8 m deep, but some may exceed 45 m in depth and 2 km in diameter.

ESKERS

Eskers are among the most remarkable of ice contact stratified drift phenomena and consist of long, narrow, sinuous ridges composed of drift. They range in height from 2 m to nearly 50 m, in breadth up to 200 m, and in length from less than 100 m to up to more than 50 km in major systems; the sides are generally steep and the crests are smooth or broadly hummocky. Most eskers seem to be deposits of glacial streams confined by walls of ice and left as ridges when the ice disappeared. The most common origin appears to have been in tunnels at the base of the glacier, during a late stage of deglaciation when the ice was thin and nearly stagnant. Others were built headward in successive segments each marked by a delta where the esker stream entered a glacial lake.

Eskers, kame terraces, kames, kettles and associated features were formed in immediate contact with wasting ice. There is another type of drift deposit which can be considered under the heading of pro-glacial deposits, that is, deposits made beyond the limits of the glacier in streams and lakes. Outwash is the term used for pro-glacial stratified drift that is stream built, or washed out, beyond the glacier itself. As a sediment, most outwash resembles any other deposit made by a stream that is heavily loaded with material. Its simplest form is that of a fan resting against the terminus of the glacier. A long, narrow body of outwash confined within a valley, which is commonly termed a valley train, will have a braided stream pattern while active deposition is in progress.

VALVES

The fine-grained bottom sediments of many glacial lakes occur in pairs of coarse and fine layers, which are believed to represent annual cycles of deposition. These alternating layers are known as valves and are the most interesting of the pro-glacial deposits which can form in glacial lakes.

Clearly as many of the pro-glacial deposits will be very similar in form to the ice contact stratified drift deposits, it is not always possible to distinguish between them, for often they grade directly into each other.

The Present-day Distribution of Ice-sheets and Glaciers

The existing world distribution of major glaciers and ice-sheets is shown for the northern hemisphere in Figure 9.3. About 10 per cent of the land surface or about 3 per cent of the earth's surface is covered by ice and the Greenland and Antarctic ice-sheets

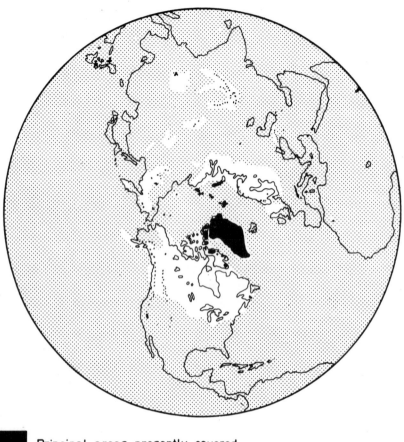

Figure 9.3 Areas covered by present and former glaciers in the northern hemisphere. (Compiled from Flint, 1958, and other sources). (After Bloom, 1969).

■ Principal areas presently covered by glacier ice

□ Principal areas covered at the last glacial maximum

together form 97 per cent of this ice-covered area, and 99 per cent of the world's land ice.

In the tropics, the natural fall of temperature with increasing altitude can compensate for the high surface temperatures; high tropical peaks can be snow-covered. Over South-East Asia the altitude of the snow-line would appear to be between 4 000 and 5 000 m, which means that there are no significant areas of ice and snow in the South-East Asian region, for most mountains do not even approach the snow-line. Snow-fields and small glaciers are found on the highest mountains of New Guinea, but at present little is known about these glaciers.

The two largest existing ice-sheets, in Greenland and Antarctica, are of interest because of the analogies they offer with the Pleistocene ice-sheets. The Greenland Ice-Sheet has an area of 1 726 400 sq km, and occupies about 80 per cent of Greenland; it has the shape of an elongated inverted dish, with a central dome reaching a maximum altitude of 3 200 to 3 300 m, and a lower southern dome. The interior gradients of the ice-sheet are very gentle (1:100 to 1:2 000) but near the margins the slopes increase to as much as 1:5. Mountains exist along the east and west coasts through which extensive glaciers flow, and some of these are up to 10 km in width. The main ice-sheet appears to occupy a lowland, which in places lies below the present sea-level and is probably in part a result of subsidence of the crust beneath the weight of the ice; the lowest point so far measured lies at −400 m, the maximum thickness of ice being 3 300 m.

Over the interior of the Greenland Ice-Sheet the climate is very cold, with mean annual temperatures of the order of $-10°C$ to $-30°C$, and it is also very dry with mean annual precipitation ranging from 500–600 mm in the south to 100–200 mm in the north. The southern coast has a more maritime climate with annual precipitation exceeding 1 000 mm. It has been estimated that the area of general ablation covers 15 to 20 per cent of the area of the ice-sheet.

In the Antarctic the ice-sheet is more than seven times greater in area than its counterpart in Greenland, for its area is about 13 100 000 sq km, most of it lying south of 70°S. The surface of the ice-sheet is not a simple dome, but is somewhat irregular and broken in places by groups of mountains. The mean altitude of the Antarctic Continent is about 2 000 m and the maximum ice thickness so far measured is 2 400 m; because of the altitude the mean annual temperatures are below freezing-point. The precipitation appears to be low, since mean annual precipitation for the continent as a whole has been deduced as being 70 to 90 mm, but for the coast belt it reaches 180 to 250 mm and falls to only 20 mm at the South Pole.

Pleistocene Glaciations

The Pleistocene glaciations marked the climax of a gradual and prolonged climatic worsening during the later Tertiary, for in the early Tertiary the climate of the earth was warmer and more uniform than at present. Temperatures fell constantly from the Eocene onwards, and in middle latitudes plants rather like those existing at present replaced tropical and sub-tropical species. In the mid-Tertiary the sub-tropical forest of the Pacific coast of North America migrated southwards through 1 600 km to south California and Mexico. Antarctica cooled rapidly during the late Tertiary, for originally a luxuriant vegetation grew in the vicinity of the Antarctic Circle and the Antarctic, like the Arctic, furnishes no evidence of glaciation before the ice age. Similarly, the climate of Australia was probably about 10°C warmer during the Miocene than at present.

Since about 90 per cent of the present glacial ice is found in the Antarctic ice-sheet, which forms one of the major controls on the climate of the southern hemisphere, the history of the Pleistocene glaciation is very much the history of this particular ice-sheet. Although it occupied a polar or near-polar position throughout the Cenozoic, Antarctica apparently did not support continent-wide ice-sheets until late Miocene time, but evidence points to the existence of relatively small complexes of mountain glaciers and highland ice-caps during Eocene and Oligocene time. By about 7 million years ago, in the late Miocene or early Pliocene, a large ice-sheet existed in West Antarctica, and by at least 4 million years ago in the middle Pliocene a major ice-sheet had formed in East Antarctica. The water removed from the oceans by the formation of these ice-sheets must have caused a world-wide lowering of sea-level by about 55 m. Since its formation, the fluctuations in the Antarctic ice-sheet have involved changes around the edge rather than changes of surface altitude.

Following the general cooling in late Tertiary times it might be expected that there would be one major ice age followed by a general warming. However, much of the original work on the Pleistocene glaciation was carried out in the Alps, and northern Europe, where it is possible to recognize four major glacial periods separated by interglacials when the ice retreated, and similar evidence has also been found in North America. Therefore it is often stated that there were four glacial periods separated by warm interglacials, but, on the other hand, there is evidence that the Antarctic ice-sheet has persisted for several million years since the late Tertiary and this suggests one glacial period with marked pulsations of the ice-sheets which affected Western Europe and North America. The latter view implies that we are still in an ice age and that the impression of several ice ages in the Alps and North America arises because these areas were ice-free during the relatively warm periods.

It is therefore necessary to think in terms of two stages in the general climatic deterioration. At the start of the Tertiary, the world was considerably warmer than at present and glaciers only occurred in the high mountain regions. Indeed, there is evidence of a world temperature maximum during the Cretaceous and early Tertiary, with mean global temperatures higher than at any time since the Carboniferous glaciations. Because of the upper limit imposed by evapotranspiration on temperature (see Chapter 7), humid tropical temperatures were probably only slightly higher than at present, but middle latitude and polar temperatures were considerably higher, leading to very weak north-south temperature gradients. Except at very high latitudes, mean annual sea-level temperatures were probably everywhere above 0°C, while in the humid tropics they perhaps reached 30°C as compared with the existing 26° to 27°C. The higher temperatures and also the extensive shallow continental seas that existed at the time, would probably cause higher rainfalls than at present, though the major arid regions would still have existed under the influence of the sub-tropical anticyclones.

The first stage of the Tertiary climatic deterioration consisted of the cooling from the early Tertiary climatic optimum to the general glacial climate of the present. This cooling was most marked in middle and high latitudes and produced the steep north-south temperature gradients which are observed today. The general cooling was accompanied by a retreat of tropical-type conditions and vegetation complexes into the present tropical zone. World climate today is therefore far from normal and probably not typical of much of the geological past of the planet. Since near tropical conditions were widespread in middle latitudes during the early and middle Tertiary, it should not cause surprise if landscapes are found to contain remnant features which appear to have formed under warmer conditions than those existing now.

If the present climate is termed a 'normal glacial climate', then there have been several periods during the last million years when the climate has become 'severe glacial'. During these severe glacial periods further extensive cooling took place in middle and high latitudes compressing the tropical zone still further, while extensive ice-sheets formed over North America and Europe. Whereas the formation of the Antarctic ice-sheet in the early stages of the glacial period lowered global sea-levels by about 55 m, the formation of further ice-sheets during the last severe glacial episode probably decreased sea-levels by an additional 150 m. These severe glacial periods form the second stage of climatic deterioration, and at least four are known from the geological record.

At no time during the known geological past have extensive snow- or ice-fields existed in lowland areas in the tropics, because the main results of the severe glacial periods in the tropics were to lower sea-levels and to modify climates. The actual extent of the ice in Asia during the last severe glacial period, usually known as the Würm or Wisconsin glaciation, is obscure. Cirques, striae and moraines demonstrate the glaciation of the highest summits of Japan, where there was a snow-line at about 2 800—3 000 m which fell northwards to 2 550 m. The glaciation, however, appears to have been slight. Taiwan had minor glaciers and a snow-line around 3 400—3 500 m. The present Himalayan and Tibetan glaciers are mere vestiges of the former giants which were extensive, for example, in the Pamirs where the Muksee Glacier was about 195 km long, while a powerful glaciation centred in the western Kuenlun Mountains had glaciers up to 100 km long extending into the Tarim Valley in Sinkiang. In New Guinea the glaciers were up to 15 km longer than the present ones and descended to about 2 000 m.

In other parts of the tropical world there is abundant evidence of lower temperatures during the Wurm glaciation. Africa had glaciers on some of its highest peaks which are located almost on the equator, i.e. on Mount Kenya (0° 12'S, alt. 5 195 m) glaciers descended to between 4 700 and 3 000 m. Traces of a former glaciation have been observed in the mountains of central America, while at the glacial maximum the lowering of the snow-line in the tropical Andes amounted to between 800 and 1 000 m.

Würm Climates

The onset of the Würm glaciation may be dated about 60 000 to 70 000 years ago, when there was a rapid fall in world mean temperatures to near the lowest levels observed in this particular glacial maximum. Planktonic Formaminifera found in tropical Atlantic Ocean bed deposits suggest that mean tropical temperatures may have decreased by about 5°C in less than 5 000 years. Evidence from New Guinea, based on palynological findings, indicates that the vegetation levels on the mountains were depressed by at least 850 m during the Würm glaciation. If an average fall of temperature with height of 0·55°C per 100 m is assumed, then this represents a fall of temperature of about 5°C. Evidence from equatorial Africa suggests a similar fall in temperature.

The cold period following the initial rapid fall in temperature probably lasted about 10 000 years up to 50 000 years ago, and was followed by a relatively warm period lasting till 30 000 years ago. A further cold period followed, lasting about 10 000 years, in which the lowest temperatures and the greatest extent of ice-sheets were attained. This particular cold phase reached its peak between 15 000 and 20 000 years ago. The end of the Würm glaciation occurred between 15 000 and 8 000 B.C. and was marked by a rapid rise of temperature, probably mostly within 2 000 or 3 000 years. Since about 3 000 B.C. there has been a slight lowering of world mean temperatures. The evidence is that temperature changes in most parts of the tropics have been small as compared with those in temperate latitudes, and their influence on lowland vegetation communities has probably been slight.

Changes in tropical rainfall distributions during the Würm glaciation were extensive, but they are difficult to date. There is evidence from Africa of considerable changes in rainfall, for example, Lake Chad had at one time a water-level up to 50 m above the present one, while fossil dunes are found up to 600 km south of the active dune fields of the contemporary Sahara. During the Würm glaciation there was a general compression of the climatic zones towards the equator, resulting in changes in rainfall regimes. Temperate latitude depression tracks were therefore moved equatorward bringing wet conditions to the poleward margins of the present-day sub-tropical deserts. Now while temperature changes were probably roughly simultaneous over most of the tropical world there is no evidence or indeed theory which suggests that rainfall changes should be simultaneous. So wetter conditions in one area could well be accompanied by drier conditions elsewhere. Many of the world's arid regions appear to have undergone a pluvial period towards the end of the Würm glaciation, and this pluvial is thought to have been at a maximum around 10 000 years ago and to have come to an end about 7 000 years ago. It corresponds not with the period of maximum glaciation but with the melting phase when the atmosphere was warmer and the climatic zones still displaced equatorward. It was during this humid period that the Mega-Chad with its high lake levels was probably formed. Earlier periods in the Würm were probably much drier and extensive dune systems formed in the Chad basin.

The evidence from South-East Asia is inconclusive, but it seems unlikely that the area was ever completely arid during the Würm, but the rainfall may have been lower, leading to a much restricted distribution of equatorial rain forest. Similarly, there is evidence of an extensive pluvial period in the deserts of Australia towards the end of the Würm glaciation.

Glacial—Eustatic and Isostatic Controls of Sea-level

Ice-sheets can create eustatic changes of sea-level which are world-wide, and also isostatic changes which are restricted to the region covered by the ice. The great continental ice-sheets were extremely thick and therefore created stresses on the earth's crust causing it to sink. The rocky floor under a major ice-sheet sinks for two different reasons, firstly because of the earth's elasticity and secondly because of the earth's plasticity. Under elastic deformation the restitution of form is perfect, but under plastic deformation, the restitution is neither perfect nor instantaneous but takes place slowly and the earth does not regain its original form. The addition of ice to an extensive ice-cap causes the whole planet to be immediately, elastically, distorted a little, but if without delay the extra ice is removed, the earth immediately takes on its original shape. In contrast, if an extensive load is applied for a prolonged period of time, the deep layers of the earth's crust will undergo plastic flow and a basin will be formed.

Suppose now that an extensive ice-sheet completes its basining of the earth's crust and then melts away. The uncovered region rises, at first by the elastic response of the earth, and then later even more because of the plastic flow of the deep rocks. The rise is greatest in the central region where the ice was thickest, so that the ultimate result is an updoming of the glacial tract; such updoming is indicated not only in northern Europe and North America but also in other glaciated regions. Changes in the level of the land surface because of variations in load are called isostatic changes and were discussed in Chapter 1.

The term 'eustatic' is applied to those changes in sea-level that are simultaneous over the entire world. They can be caused by changes in the volume of the ocean basins, or by changes in the amount of water temporarily abstracted from the sea by glaciers and ice-sheets. If a warm climate turns into a cold glacial climate the precipitation over large areas changes from rain into snow. The snow becomes locked up in ice-sheets and does not form river-flow back to the oceans, thus resulting in a lowering of sea-level.

The melting of the ice-sheets between about 12 000 and 9 000 years ago caused a eustatic rise of sea-level of up to 150 m. The lowest sea-levels probably occurred about 15 000 years ago, with a very rapid rise in sea-level, corresponding to the melting of the ice-sheets, between 12 000 and 8 000 years ago, present-day sea-levels being almost reached about 6 500 years ago. During the last 6 500 years there has been a further slow rise in sea-level of about 6 m.

One general result of the eustatic fall in sea-level during the Würm glaciation was that shallow seas became dry, thus extending the length of rivers such as those of Borneo, Sumatra and the Malay Peninsula across the Sunda shelf. World-wide emergence of land converted many straits into dry land and this affected the migration of varieties of plants and animals. For example, in general the list of living animal species found in Sumatra is almost identical with the list for Borneo. On the other hand, the land animals of Celebes, though relatively much nearer to Borneo, are much less related. The likeness of faunas in Sumatra and Borneo is explained by the fact that during the Würm glacial maximum, these two islands were connected by the broad Sunda land-bridge. No such land-bridge between Borneo and Celebes was possible, because the sea between them is too deep. Among other temporary land-bridges were the ones joining Tasmania to Australia and Ceylon (Sri Lanka) to India.

There is no evidence that since the end of the Würm glaciation world-wide sea-levels have been higher than at present. Local tectonic changes may create an impression that there has been a recent fall in sea-level, but tectonically stable areas only show rising ocean levels. Because of sea-floor spreading (see Chapter 2), the ocean basins are widening at rates of up to 16 cm per year. Even if a much lower rate is allowed (about 10 cm per year), a substantial volume of new ocean basin has been created during the last 100 000 years, probably enough to accommodate about 6 per cent of the returning melt-water from the Würm glaciation, and so post-glacial shorelines could be almost 8 m lower than the corresponding inter-glacial shorelines of 100 000 years ago. Indeed, the late Cenozoic appears to have been marked by rapid continental drift, vertical motion of continents and the sea-floor, and also by intense volcanism. The

volume of the ocean basins probably increased, but the volcanism caused the actual volume of sea-water to also increase by adding additional water-vapour to the atmosphere. It is not therefore possible to assume that non-glacial sea-levels have been constant, as indeed the long-term trend may be for falling sea-levels, and on this gradual fall are superimposed the various eustatic changes. Zeuner (1959) suggested that sea-levels progressively subsided about 100 m during the Quaternary, and upon this lowering was superimposed a number of fluctuations that were due to ice-sheets forming and melting.

The various actual and inferred changes in sea-level are important for the Sunda shelf region because it is tectonically relatively stable. In particular there are widespread deposits in the Malay Peninsula known as the Old Alluvium which are mainly of early or middle Pleistocene and which appear to indicate a sea-level above the present one. It must not be assumed that all such deposits show past sea-levels since they may have formed in river valleys or the higher tributaries of river systems. They do however suggest sea-levels somewhat above those of the present time and this can only be explained by a general fall in levels over the last million years or so. The exact history of Pleistocene sea-levels in the Sunda area is far from clear, and must wait for further dating of ancient shorelines and more information on changes in the volume of the ocean basins.

FURTHER READING

Bloom, A.L. (1969). *The Surface of the Earth* (Prentice-Hall, Inc., New Jersey).

Butser, K.W. (1972). *Environment and Archaeology: An Ecological Approach to Prehistory* (Methuen).

Charlesworth, J.T. (1957). *The Quaternary Era* (Arnold, London).

Flint, R.F. (1957). *Glacial and Pleistocene Geology* (Wiley, New York).

Zeuner, F.E. (1959). *The Pleistocene Period* (Hutchinson, London).

PART E

The Chemistry and Biology of the Tropical Environment

Limestone Caves
Top Left Man has used limestone caves for many thousands of years and remains such as these spirit-ship figures which are carrying souls to the afterworld, may sometimes be found.
Top Right Inside a limestone cave.
Bottom The entrance to Batu Caves, which contain a Hindu Shrine dedicated to Lord Subramaniam.

10 Matter in the Tropical Environment

The Nature of Matter

ALL matter observed on earth is composed of exceedingly small basic building blocks called atoms, each of which consists of a small central nucleus, with a positive charge, surrounded by a cloud of negatively charged electrons. Each electron is in orbit around the nucleus, and the diameter of the nucleus is more than 10 000 times smaller than that of a typical electron orbit. However, the bulk of the mass of the atom is contained in the nucleus, which consists of positively-charged particles called protons, and particles with a similar mass (quantity of matter) to that of the proton, but without an electric charge, called neutrons. The positive electric charge of the proton is exactly equal to, but opposite to that of the electron, while its mass is about 1 850 times that of the electron. In the normal complete atom, the number of electrons is exactly equal to the number of protons, and it is the number of positive charges in the nucleus (i.e. protons) that defines the atomic number of the atom. In contrast, atomic mass is the number of both protons and neutrons in the nucleus, and since neutrons carry no electric charge it is possible, because of variations in the number of neutrons, to have a nucleus which has the same atomic number as another nucleus but a different atomic mass. Such variations in atomic mass among atoms which have the same atomic number give rise to what are known as isotopes. So an isotope is an atomic nucleus which has the same atomic number as another nucleus but a different atomic mass.

If an atom should for some reason lose one of its negatively-charged electrons, it will become positively charged and form a positive ion, and similarly if it were to gain an electron it would become a negative ion. Since the number of electrons in an atom is not constant, the type of atom is determined by the number of positive charges in the very stable nucleus.

So hydrogen has one proton in its nucleus and normally one electron, chlorine seventeen protons and normally seventeen electrons, and lead forty-six protons and normally forty-six electrons. The property of the atom to gain or lose electrons controls its chemical properties, and atoms all with the same atomic number form a single element. So an element can be defined as one of a number (about 103) of simple substances in terms of which the chemical composition of all other substances may be expressed. When positive ions of one element meet negative ions of another element they are attracted and form a new substance which is a compound of the two original elements, and in this general manner it is possible to build a whole range of new compounds from the original elements; combinations of atoms in compounds and elements are known as molecules. A very simple compound which is formed in this manner is a salt (NaCl), which is the product of a reaction between an acid such as hydrochloric acid (HCl) and a base such as sodium hydroxide (NaOH).

$$NaOH + HCl \rightarrow NaCl + H_2O$$

The atomic weight of an atom is its relative mass on an arbitrary scale, and similarly the molecular weight of a compound is the sum of the atomic weights of all the atoms represented in the chemical formula of the compound. When the number representing the molecular weight of a substance is given a value in grammes, then this quantity of the substance is called a gramme molecular weight, or in abbreviated form, a mole. Provided that the same number of molecules is involved, the ratios of the molecular weights of compounds will always be the same. Hence a mole of any one substance contains the same number of molecules as a mole of any other substance.

The Solvent Power of Water

A chemical property of water which makes life possible as it is observed on earth, is its power to dissolve many substances. Of all liquids, water is probably the best solvent, since most substances are at least slightly soluble in water, and many dissolve in it to a considerable extent.

Sodium chloride (common salt) is a solid at normal temperatures and only melts if it is heated to about 800°C, but if salt is mixed with water it dissolves to give a homogeneous solution. Although the original salt and water had different properties, the solution of salt in water has uniform properties throughout. This is where a solution differs from a mixture, e.g. salt and sand, which does not have uniform properties and can be divided into its individual components. To separate a solution of salt in water it is necessary to boil off the water leaving the salt behind as a residue.

Crystals of sodium chloride consist of regular arrays of positively-charged sodium atoms and negatively-charged chlorine atoms, which are held in place by their strong electrical attractions. The positive and negative ions are so firmly held in position that solid sodium chloride is a good electrical insulator. When sodium chloride dissolves in water the crystal structure is destroyed and the ions become mobile. If charged electrodes are introduced into the solution, the positive sodium ions will be attracted to the negative electrode and the negative chlorine ions to the positive electrode, and so an electric current will flow through the salt solution. All chemical salts become dissociated into positive and negative ions when they dissolve in water and these ions become mobile as independent entities.

Pure water consists of water molecules, but even in pure water some of the water molecules are dissociated resulting in positively-charged hydrogen ions and negatively-charged hydroxide ions.

$$H_2O \rightarrow H^+ + OH^-$$

Under equilibrium conditions the rates of dissociation and combination are equal, and so in pure water the concentrations of H^+ and OH^- ions are equal at 10^{-7} mole kg^{-1}. Hydrogen and hydroxide ions are very important, for, in combination with negative and positive ions, respectively, they form acids or bases. Examples are hydrochloric acid (HCl) which consists of a hydrogen ion plus a chloride ion and the base sodium hydroxide (NaOH) which is a combination of a sodium and a hydroxide ion. If equal amounts of HCl and NaOH are mixed, they react to form a neutral solution of sodium chloride, since the hydrogen ions combine with the hydroxide ions to form water.

$$NaOH + HCl \rightarrow Na^+ + Cl^- + H_2O$$

A neutral solution of sodium chloride contains 10^{-7} mole kg^{-1} each of hydrogen and hydroxide ions. If a solution contains more hydrogen ions than hydroxide ions it becomes acidic, while an excess of hydroxide ions produces a basic reaction, but in any given solution the product of the two ion concentrations is always 10^{-14}.

The pH of a solution is the negative logarithm of the hydrogen-ion concentration, and is thus a measure of the acidity of the solution. If the solution is neutral, the hydrogen-ion concentration is 10^{-7} and the pH $= -\log 10^{-7} = 7$. So a neutral solution has a pH of 7, but if it is acidic the pH will be less than 7, while if it is basic it is more than 7. Typical pH values observed in the natural environment range from 9 for alkali soils through about 8 for sea-water, to about 4 or 5 for acid soils and peat water.

The solubility of many substances varies greatly with the pH of the water. For example, the solubility of iron is about 100 000 times greater at pH 6 than at pH 8·5. At pH values below 4, alumina is more soluble than silica, but between pH 5 and 9 alumina is virtually insoluble but silica becomes increasingly soluble.

In a neutral solution the total electric charge of the negative ions equals that of the positive ions. Now while the ions of sodium, chloride and many other elements are singly charged, others like calcium (Ca^{++}) and magnesium (Mg^{++}) have ions which are doubly charged. It is therefore convenient to express the concentrations of ions in terms of equivalents, where an equivalent is the amount of an ion required to equal the charge of a gramme-molecular weight of a singly-charged ion. So for a singly-charged ion the number of equivalents is equal to the number of moles, while for a doubly-charged ion, one mole is equal to two equivalents. In a normal solution of salts, the sum of the equivalents of the negative ions must equal the sum of the equivalents of the positive ions, so that the solution remains electrically neutral.

Chemical Cycles

Many elements on the earth's surface are involved in chemical cycles which may be considered in terms of cascading systems with several reservoirs or stores with transfers between each of the reservoirs. The chemical reservoirs may be the atmosphere, oceans, solid rocks, soil or plants, and the transfers result from weathering and erosion, plant growth and decay, etc. The solid earth forms a chemical reservoir which is extremely large when compared with the others, so the nature of chemical cycles can best be illustrated by examples from the atmosphere and oceans.

Careful observation reveals that, with the possible exception of the noble or inert gases (He, Ne, Ar, Kr, etc.) all atmospheric gases go through a chemical cycle, with varying residence times in the atmosphere. Studies of these cycles suggest that the abundance of many of the elements in the atmosphere and oceans cannot be explained only by the weathering of igneous rocks. They must have accumulated partly because of the release of gases from volcanoes and hot springs, since volcanic gases contain elements which are new to the atmosphere, originating as they do deep in the earth's interior. The amount of a particular gas in the atmosphere could therefore be controlled either by a chemical cycle or by exhalation from the earth's interior.

If the mass of an atmospheric gas, including the part dissolved in the oceans (the mobile fraction), is greater than the amount of the same constituent deposited in sedimentary rocks, then most of the constituent has remained in the atmosphere and it may be termed an accumulative gas. If, on the other hand, the mass of the atmospheric gas is less than that of the same constituent contained in sediments or being released from volcanoes, then the amount remaining in the atmosphere is the result of equilibrium conditions in a chemical cycle.

Carbon dioxide is a good example of an equilibrium gas. Only about one-thousandth of all the carbon dioxide released from the earth's interior has remained in the atmosphere and oceans as a mobile gas, with only about one-sixtieth of this total being present in the atmosphere. The carbon dioxide released by volcanic activity reacts through the process of weathering with the calcium contained in igneous rocks to form soluble bicarbonates which are carried by the rivers into the oceans. The oceans are nearly saturated with bicarbonates and as a consequence of this, the additions from rivers are precipitated as carbonates in the form of limestones, dolomites, etc., mostly by the action of living organisms. The carbon dioxide-carbonate cycle is of great importance and is considered separately in a later section of this chapter.

Neon, argon, krypton and xenon, together with nitrogen, are examples of accumulative gases. Oxygen is an interesting gas because it does not fit clearly into the two general categories of accumulative and non-accumulative gases. Most of the oxygen in the atmosphere has been produced by photosynthesis in plants, but part of this oxygen is consumed by the oxidation of minerals, so the oxygen in the atmosphere is probably the result of both accumulation and a geochemical cycle.

Ocean Water and River Water

The hydrologic cycle is the name given to the process by which water is evaporated from the oceans, recondensed in the atmosphere to form rain, which in part falls on the land, is gathered in rivers and returns to the sea. As the rain-water seeps through and over the ground, it dissolves some of the mineral matter and carries it to the sea. Thus the rivers are continually adding fresh minerals to the sea, and it might be expected that sea-water would have a composition which is approximately that of concentrated river water. However, analysis of water from a large number of streams (see Table 10.1) indicates that river water is mainly a solution of calcium bicarbonate, with considerably smaller amounts of chlorides and sulphates of sodium, potassium and magnesium, whereas in contrast, the dominant salt in sea-water is sodium chloride, with magnesium and sulphate ions being next in importance, there being only a trace of calcium bicarbonate.

Tiny crystals of sea salts are formed in the atmosphere by the evaporation of spray and some of these are carried over the continents by the winds. Precipitation washes these small particles out of the atmosphere and they are also deposited in dry weather on plants and the soil. In wet weather they are carried into the rivers and back to the sea. This process results in a recycling of sea salt, and the salts so derived are called cyclic salts. The concentration of cyclic salts in river water is high just downwind of the sea

TABLE 10.1

COMPOSITION OF RIVER AND SEA-WATER
Concentration in milliequivalents per kilogram
(See section on Solvent Power of Water)

	Average River Water	Average Sea Water
Sodium (Na^+)	0·27	468·0
Potassium (K^+)	0·06	10·0
Magnesium (Mg^{++})	0·34	107·0
Calcium (Ca^{++})	0·75	20·0
Total cations	1·42	605·0
Chloride (Cl^-)	0·22	546·5
Bicarbonate (HCO_3^-)	0·96	2·3
Sulphate (SO_4^{--})	0·24	56·2
Total anions	1·42	605·0

Source. (After P.K. Weyl (1970). *Oceanography, an Introduction to the Marine Environment* (Wiley, New York).)

coast and decreases inland. To obtain the true concentration of salts in river water derived from rock weathering, it is necessary to subtract the cyclic salt contribution from the total mineral content in solution. This is not too difficult since weathered material usually yields little chloride ion, and it is therefore possible to assume that all the chloride ions in river water come from cyclic salts.

Calculations show that it would take 44 000 years for the rivers to add to the oceans a mass of water equal to the content of the oceans. The time that it would take to double the water content of the oceans assuming that rivers continued to flow at their present rate and that there is no loss of water from the oceans is called the residence time of water in the oceans. Similar calculations can be made for the salt content of the oceans, and it is found that the average residence time for total salt is 22 000 000 years. The residence times for individual ions vary considerably ranging from 260 000 000 years for sodium, 16 000 000 years for magnesium, 1 200 000 years for calcium and 110 000 for bicarbonate. Geological evidence suggests that as the oceans are at least 1 000 000 000 years old, all the above residue times are short compared with the age of the ocean, indicating that the oceans do not just accumulate salts from rivers but can dispose of them as well. This disposal takes the form of precipitation of chemicals

from the ocean water to form new deposits. Thus the constituents dissolved in the oceans are in a steady-state balance, though rivers supply a very varied selection of ions.

Weathering

Weathering is the adjustment that the minerals of rocks make in response to the presence of air, water, and other changed conditions at the surface of the earth. The physical and chemical environment at the surface differs from that inside the earth, so changes occur when rocks from deep in the earth become exposed at the surface. Some of the products resulting from these changes accumulate on the ground, forming a soil, while others are removed by erosion as small particles or in solution. Weathering processes can be classified into two categories, corresponding to physical and chemical changes in the rock materials.

PHYSICAL WEATHERING

This includes all those processes in which large fragments of rocks and minerals are broken into smaller fragments without changing the chemical composition of the material. Four agents are primarily responsible for this particular type of weathering; they are the freezing of water, the growth of crystals, organic activity and abrasion. Water expands by about 9 per cent when it freezes, and if this occurs in a crack in a rock the expansive force will be sufficient to fracture the rock. Obviously this particular process is most effective in cold climates where there is repeated freezing and thawing. Growth of large crystals due to chemical changes within a rock can have the same effect, and cause rock fractures. The chemical changes are usually caused by chemical weathering and are especially important in rocks composed of silicate minerals, since most silicate minerals weather chemically to form clay minerals which occupy a larger volume than the original minerals. Similarly plant roots can grow into underlying rock banks, causing the rocks to split. Other agents of physical weathering include earthworms which pass soil through their digestive tracts, breaking and grinding mineral grains in their stomach. Small particles are moved by the wind, running water and ice, and frequent collisions cause them to break into smaller fragments. This type of weathering can be important in arid and semi-arid areas.

CHEMICAL WEATHERING

This includes all processes in the formation of new minerals which are in equilibrium with the environment at the surface of the earth. Rock minerals forming deep in the earth exist in an environment where there are high temperatures and pressures, where all the oxygen is chemically combined, carbon dioxide is rare or absent, and there is little or no liquid water. In contrast, at the surface, temperatures and pressures are low, there is normally abundant liquid water, uncombined oxygen exists in the atmosphere, and also carbon dioxide gas exists both in the air and dissolved in rain-water. It would seem likely that the most unstable rocks will be those formed under conditions least like those at the surface, for instance, certain igneous and metamorphic rocks, while sedimentary rocks such as shales and sandstones will be among the most stable because their minerals have already been subject to some weathering. The important processes of chemical weathering can be grouped under the headings of hydration, hydrolysis, solution of metallic ions by natural acids, solution of soluble minerals in water, oxidation and reduction.

Hydration. This is the absorption of water by a mineral, where absorption is the penetration of water into the surface layers of a solid. During this process, the H^+ and OH^- ions from water penetrate between the plates of the crystal lattice of the mineral, making it porous and subject to further weathering. A good example is the formation of yellow limonite from red haematite:

$$2Fe_2O_3 + 3H_2O \rightleftharpoons 2Fe_2O_3 . 3H_2O$$

The process is reversible since dehydration will produce haematite again.

Hydrolysis. This is the reaction of water with a mineral, and is thus closely associated with hydration. A new mineral is created which contains water, a good example being the formation of clay from feldspar:

$$K . Al . Si_3O_8 \;+\; H_2O \rightarrow Al_2 Si_2O_5 (OH)_2 \;+$$

Potassium feldspar   Water   Kaolinite clay remains

K and Si Oxides
dissolved in water

Hydrolysis is an important chemical weathering process, since the decay of coarsely-grained rocks such as granite, even in hot desert environments, may be attributed more to the hydrolysis of the feldspars than to any other action.

Oxidation. This is the addition of oxygen to a mineral, often by the action of aerobic bacteria. Atmospheric oxygen forms metallic oxides from silicate or sulphide minerals, which can then be removed by other weathering processes.

$$2Fe_2 SiO_4 + O_2 \;\longrightarrow\; 2Fe_2O_3 + 2Si O_2$$

Olivine   From   Haematite   Dissolved
           Air   remains   in water

Reduction. This is the release of oxygen from compounds, often by the action of anaerobic bacteria. Reduced conditions are found in soils or regoliths with slow water movement or waterlogged conditions.

Solution of Soluble Minerals in Water. Sediments that form by the evaporation of water in an arid climate, such as rock salt and gypsum, are readily dissolved by water in humid regions.

Solution of Metallic Ions by Natural Acids. Carbonic acid, formed when atmospheric carbon dioxide dissolves in rain, is one of the most commonly-occurring natural acids.

$$H_2O + CO_2 \rightleftharpoons H_2CO_3$$

Carbonic acid removes sodium, potassium, calcium and magnesium from many types of rocks to form a bicarbonate salt in aqueous solution. A very common reaction is that between limestone and carbonic acid to produce calcium bicarbonate.

$$H_2CO_3 + CaCO_3 \;\longrightarrow\; Ca (HCO_3)_2$$

Carbonic   Limestone   Calcium Bicarbonate
acid   in solution

Sulphuric and various organic acids occurring in nature are also important weathering agents.

COMMON WEATHERING PRODUCTS

As silicon, sodium, potassium, magnesium and aluminium are among the common rock-forming elements, it is useful to note their weathering products.

Silicon. This element occurs in two forms, as quartz and as a constituent of complex silicate minerals. Quartz is essentially insoluble under ordinary weathering conditions, so if quartz occurs in a rock it is left as a residue of quartz sand after the more soluble minerals have disappeared. This in turn may be eroded and then deposited elsewhere to form sandstone. Silicon also occurs in certain rock crystals, and after the metallic ions are dissolved, it goes into aqueous solution as silicon dioxide.

$$NaAlSi_3O_8 + H_2CO_3 \longrightarrow Al_2Si_2O_5(OH)_2 +$$

Sodium   Carbonic   Kaolinite clay
feldspar   acid   remains

$$NaHCO_3 + SiO_2$$

Sodium   in solution
bicarbonate
in solution

Sodium, Potassium, Calcium, Magnesium. These elements dissolve with varying degrees of ease, and are transported toward the sea in bicarbonate solution as in the examples for potassium and sodium feldspar.
Aluminium. Since aluminium does not dissolve during weathering, it accumulates in the soil, usually as a constituent of clay minerals. The clays are further decomposed in tropical climates and the aluminium remains in the soil as aluminium oxide (bauxite).

$$Al_2Si_2O_5(OH)_2 \longrightarrow Al_2O_3 \, nH_2O + SiO_2$$

Kaolinite clay   Bauxite in soil   in solution

WEATHERING AND CLIMATIC CONDITIONS

Climate determines the temperature and moisture regimes under which weathering takes place. In arid regions, mechanical weathering is dominant and small particles are produced without much alteration in composition, while in humid climates, chemical weathering of minerals and the synthesis of clays are important. The speed of chemical reactions is also greatly increased by a rise of temperature, and an increase of $10°C$ can double or even treble the reaction rate. Increases in temperature also alter the relative mobility of minerals, so that while quartz is highly resistant to weathering in temperate climates, fine-grained quartz particles are relatively easily weathered under tropical conditions. In contrast, iron and aluminium hydrous oxides are relatively more resistant under tropical conditions and therefore they tend to accumulate in tropical soils. Humid tropical climates allow a high rate of organic matter production and decay, so the supply of organic acids to take part in chemical weathering is high in tropical forests and low in temperate ones.

The Carbon Dioxide–Calcium Carbonate System

Both carbon dioxide and calcium carbonate play important parts in the surface environment Atmospheric carbon dioxide absorbs infra-red radiation thereby affecting the heat balance of the earth, and it also forms the carbon source for all plants. Some of this organic carbon becomes buried in sediments and may yield fossil fuels such as coal, petroleum and natural gas. When carbon dioxide dissolves in water, it is hydrated to H_2CO_3 and by dissociation gives rise to the bicarbonate ion, HCO_3^- and the carbonate ion, CO_3^{--} The carbonate ion, when combined with ions of the alkaline earths, primarily calcium and magnesium, forms large deposits of carbonate rock such as limestone and dolomite. If rain-water in which atmospheric carbon dioxide dissolves falls on limestone rocks, it reacts with them to produce calcium and bicarbonate ions in solution.

$$CO_2 + H_2O \longrightarrow H_2CO_3$$

$$H_2CO_3 + CaCO_3 \longrightarrow Ca^{++} \; 2HCO_3^-$$

In extensive limestone areas this type of chemical reaction can produce a type of landscape called karst topography after the Karst district of Yugoslavia where it is well developed. Limestones can consist almost completely of carbonates of calcium and magnesium; for instance the analysis of some typical Sarawak limestones showed that they were 97 per cent Ca CO_3, $1\cdot3$ per cent Mg CO_3 and that the remaining $1\cdot7$ per cent was insoluble residue. Acid rain and river water will therefore be able to completely dissolve the limestone rock leaving little in the way of mineral residue.

In the geomorphic development of soluble rocks, water acts not only at the surface but also at subterranean levels. On the surface of an extensively exposed bed of limestone the simplest solution phenomena are small swallow holes which rapidly coalesce and give rise to small channels which penetrate deeper and deeper to form a clint and grike landscape. A clint is a bare, level surface developed on horizontal beds of limestone, while a grike is a vertical fissure formed by solution along joints. Often the greater part of a bed of limestone will be removed leaving only isolated hills.

At depth, water dissolves the limestone as it follows joints and fissures, and thus it erodes the limestone underground forming a subterranean network of shafts, caves, siphons and passages. When the limestone is very thick, surface rivers carve deep canyons or plunge underground by way of sink holes only to reappear later at a lower level. The development of

Limestone rubble and organic litter commonly mixed
with clay, sand and limonitic gravel

3 m

Organic soil with matted roots

Limestone

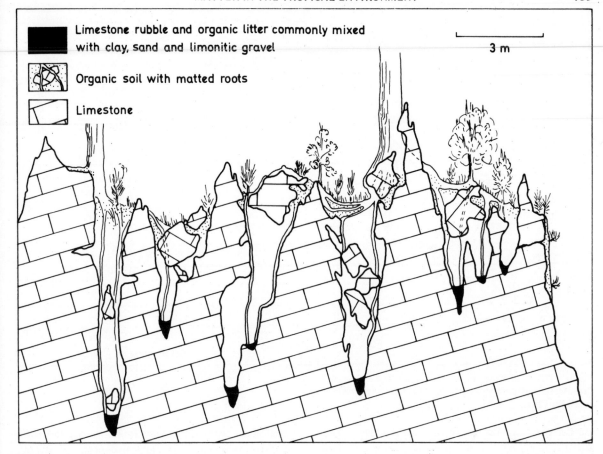

Figure 10.1 Irregular surface of a limestone mountain.
(After Wilford and Wall, 1965).

karst will thus deprive the landscape of its surface streams, which will only exist in deep gorges or underground. Communication between the surface run-off and the underground drainage system is maintained by swallow holes, which, if they are particularly large, are known as dolinas.

Some good examples of karst topography are found in Sarawak, where there are two contrasting types of limestone landscape. Most striking are groups and ranges of isolated, steep-sided, rugged hills and mountains, while less obvious are the alluvial and, in many places, swampy areas underlain by limestone, from which outcrops of bed rock project as pinnacles. Most of the limestone mountains are cliffed or have boulder-strewn flanks, the surface (Figure 10.1) being irregular and broken by numerous depressions. Cave systems, some with passages as high as 60 m and as wide as 90 m, are common in limestone hills. Caves

(Figure 10.2) are essentially horizontal and by far the greatest number occur within a few tens of metres of the water-table, those at the base of cliffs containing small streams.

Rain-water in Sarawak is generally acid, with a range in pH of 4·9 to 6·0, according to Wilford and Wall (1965), who also analysed stream water in the headwater of the Tubeh River, Bau region. The samples were collected when the rivers were high, some 8 hours after 3 days of wet weather during which about 125 mm of rain fell. The accompanying Figure 10.3 and Table 10.2 show that the river rises among the ridges of basalt and is more than half-saturated with calcium and magnesium before reaching the main limestone hills. In contrast, small streams emerging from the foot of the limestone hills had completely saturated water.

Sea-water normally contains calcium ions which

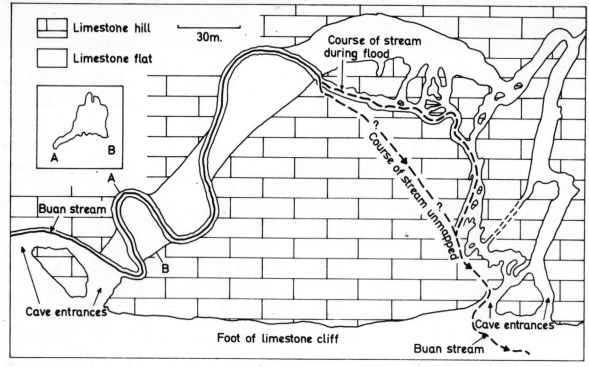

Figure 10.2 Buan Cave, near Paku, Bau Region, Sarawak. Cross-section of cave shown along line A-B. (After Wilford and Wall, 1965).

are derived from the continents via the rivers. They are the result of the solution of pre-existing limestones and the action of humic acids formed during the decomposition of organic matter on the continental surface. Within the sea, carbonates in general and limestones in particular, are precipitated by biochemical activity. The nature of the precipitation process is complex, but it can be summarized as

$$Ca(HCO_3)_2 \rightleftharpoons CaCO_3\downarrow + H_2O + CO_2\uparrow$$

where the calcium carbonate is precipitated and the carbon dioxide is liberated to the atmosphere. The exact type of rock deposited depends on the environmental conditions within the sea. Thus on the continental shelf, between depths of 0 and 200 m and exceptionally at greater depths, conditions are favourable for the deposition of limestone, especially in warm water between 10°C and 30°C. In zones of clear water, rich in nutrients, and particularly favourable to living organisms, there occur bioherms and biostromes, while other clastic carbonate deposits such as oolites may also be formed. Bioherms are deposited from the remains of various organisms ac-

cording to temperature and environment, and correspond closely to the ecological concept of a 'reef'. They are generally without bedding and consist of mounds or domes rising above the contemporaneous beds surrounding them. In contrast, biostromes are composed of stratified beds containing whole organisms which lived *in situ*. In the continental slope and deep sea regions between 200 m and 6 000 m in depth in warm climates and in open oxygenated basins, calcareous oozes of various types are formed.

Erosion

Erosion is the term used for the loosening and transport of rock and mineral debris under the influence of gravity alone or by water, wind and ice; weathering is normally the first stage of this process. Landscapes consist very largely of slopes which are formed by the process of erosion, and the removal of mass from a landscape can be viewed in terms of a cascading sediment system, with the various components of the slope being the physical expression of the system. Thus screes and debris represent stores,

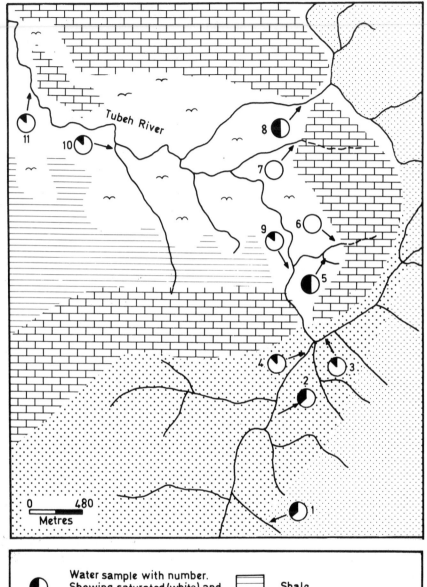

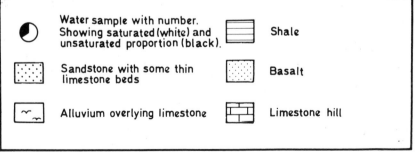

Figure 10.3 Results of stream water analyses in the headwaters of the Tubeh River, Bau region, Sarawak. The samples were collected when the rivers were high, some 8 hours after 3 days of wet weather during which about 125 mm of rain fell. (After Wilford and Wall, 1965).

and movements of material represent transfers, within the system. If the system is assumed to be in a steady state, then the elements of the slope complex which represent the various transfers and stores within the system will keep their identity over long periods of time. Steady-state conditions imply a stable environ-

TABLE 10.2

CALCIUM AND MAGNESIUM CONTENTS OF STREAM-WATER FROM THE HEADWATERS
OF THE TUBEH RIVER, BAU REGION

Sample number[1]	Calcium ppm	Magnesium ppm	Additional calcium required to reach saturation[2] ppm	pH
1	6	4	4	7·3
2	7	1	6	6·9
3	11	2	3	7·2
4	12	2	3	7·1
5	8	trace	8	6·3
6	42	trace	nil	7·4
7	27	1	nil	7·5
8	5	2	5	7·4
9	13	2	3	7·2
10	41	1	3	7·1
11	22	1	2	7·2
12	37	4	nil	7·0

[1] Figure 10.3 shows the location of Samples 1 to 11. Sample 12 was from a 15-cm diameter hollow in a limestone pinnacle choked with leaves and twigs and with a layer of rotted organic matter at the base.

[2] The additional calcium required to reach saturation was determined by shaking 100 mls of each water sample with 1 g of pure dry calcium carbonate powder for 4 hours at about 28°C and then re-determining the calcium in solution with E.D.T.A. (After Wilford and Wall, 1965.)

ment but not necessarily continuous uplift, since the slope complexes may remain stable in a gradually lowering landscape.

Slopes can be classified according to the nature of the controlling processes. So steep slopes, where debris is removed by the effect of gravity as soon as it is produced, are known as bare-rock or gravity-controlled slopes. The debris produced by weathering on gentler slopes passes downslope by mass movement forming a thin coarse mantle on a debris-controlled slope. On still gentler slopes the debris is finer and is moved by slope wash, and these are known as wash-controlled slopes.

Gravity-controlled slopes are particularly common in deserts where there are outcrops of resistant rocks. These bare-rock slopes usually occur near the tops of composite hillsides or where there has been deep incision by channels, and their exact nature depends very much on the character of the local bed rock. So relatively unresistant rocks tend to form rounded upper slopes, while more resistant ones form rectilinear faces. A rather common type of free face is composed of vertical columns or plates, parts of which

break away and add debris to the hillside below, forming a scree or debris-controlled slope.

Unloading is an important phenomenon observed on gravity-controlled slopes. Many rocks crystallize deep in the earth's interior under conditions of high pressure, and in particular this applies to granitic rocks. As erosion removes the superincumbent load, hydrostatic pressures are decreased and the relief of pressure is expressed in a series of fractures disposed tangentially to the direction of stress, that is to say parallel to the land surface. Further cracks develop inwards from the surface, forming plates or blocks which break away leaving smooth curved rock faces exposed at the surface. This type of phenomenon is found in a wide variety of rock types and in many different climates. The resulting characteristically-rounded mountains are known as domed inselbergs and are typical of many tropical arid and semi-arid landscapes.

An inselberg is an isolated or island mountain, and its essential feature is its abrupt rise from the surrounding plains, whether the latter be caused by erosion or deposition. Inselbergs of granitic composition may take the form of ridges and ranges, such as the Harts Range in Central Australia, where the uplands have been moulded into two basic shapes. The first is the angular castellated type known as the castle koppie and the second the domed inselberg or bornhardt. Many inselbergs, whether domed or castellated, give the impression of being round or oval in plan, but detailed mapping reveals that they are in reality rectangular or rhomboidal, or are composites of several blocks of these shapes. Many inselberg landscapes display bed rock which, though petrologically or lithologically uniform, is strongly varied structurally, and the principal reason for the contrast in weathering between the uplands and plains appears to be the variation in joint pattern. Thus there is a scarcity of joints on the inselbergs but numerous open joints in the plain areas. So inselbergs appear to owe their origin to structural variations in the bed rock, and mostly consist of monolithic masses of granite or similar rock.

Debris-covered slopes are probably dominated by the mass movement of material and also by the individual movement of large blocks moving rather more rapidly down the slope. Slopes can vary widely in inclination, but often they appear to change character at about 28°, and this is particularly so in desert regions. Slopes steeper than 28° are unstable, while gentler slopes are stable. It is interesting to note that the stability angle of water-saturated debris is about 26°, so even in desert regions, rare storms may be important in controlling slope angles.

Slopewash processes are important in fine-grained and incohesive materials, such as weathered mantles, and at the foot of debris-controlled slopes in more resistant materials. Slopewash processes include the action of both raindrops and unconcentrated wash. When rain falls on a wash-controlled slope, the upper part of the slope transmits run-off as a thin sheet in which the flow is laminar. Further down the slope rain falling on the surface considerably adds to the efficacy of the water as an eroding agent, and here the flow can be described as disturbed. Near the foot of the slope the flow may become truly turbulent. Unconcentrated wash is effective as an erosion agent and also sorts the debris produced by rock weathering. In arid Australia it has been found that wash is capable of moving particles up to about 5 mm.

When rain falls on a bare soil the initial raindrops break down the soil aggregates by impact, detaching soil particles and forming muddy suspensions. Further raindrops may cause splashing, which tends to displace the muddy suspensions with a preferential downslope movement. Where the soil surface varies in resistance, differential raindrop erosion may form earth pillars, and these are observed in areas with bare soil surfaces such as semi-arid regions.

In tropical rain forest only a small proportion of the raindrops fall directly to the ground except at canopy openings. Most of them are intercepted by the leaves in the canopy, and form waterdrops which fall from the tips of the leaves. Because of the great height of the canopy in tropical rain forest, many of the waterdrops are falling at near the terminal velocity for air when they reach the ground and are very effective agents of erosion. Waterdrop erosion is thus more important under tropical rain forest than under temperate forest because of the higher upper storey, the greater rainfall, the sparse shrub layer, and the less abundant leaf litter. In temperate regions, mechanical erosion is slower under natural forest than under grassland, because of the well-protected soil surface, but this does not necessarily apply in the tropics. In mature primary rain forest, canopy openings commonly occur, the shrub layer is sparse owing to the lack of light, and the rapid vegetative decay allows little leaf litter to form on the soil surface. Heavy rain is not necessary for raindrop erosion, because light rain or drizzle will re-form in the rain forest canopy into waterdrops which may be of large size.

Slopewash has been observed under mature primary rain forest in Papua on nearly all slopes above 5° and on clayey silt soils on slopes down to 3°. It performs three functions in primary rain forest which are the removal of leaf litter, the removal of muddy suspensions formed by raindrop and waterdrop erosion, and the erosion of the soil surface. During prolonged and intense rainstorms, rain-water sweeps away loose leaf litter by direct hydraulic action and re-deposits it downslope. Raindrop and waterdrop erosion involves a very wide range of soil particle sizes, since drops of large size and high impact velocity can move particles of up to 10 mm in diameter. The smaller of these particles are carried away as muddy suspensions and deposited at small obstructions such as root barriers. The water films are capable not only of sweeping away loose particles but also locally of actually detaching particles when the flow becomes concentrated. Erosion, transportation and deposition by rivers are considered in Chapter 8.

Wind can be an active agent of erosion and is capable of producing distinctive landforms, but an analysis of landscapes shows that landforms produced by wind action are of relatively minor importance compared with those produced by running water. An important form of wind erosion is deflation, that is the lifting and entrainment of loose particles of clay and silt sizes by turbulent eddies in the air. Boundary layers and atmospheric turbulence were considered in Chapter 5. The process is similar to the suspension of fine sediment in stream flow, in that grains are carried up by vertical currents which exceed the settling velocities of the grains in still air. Deflation normally occurs where clays and silts in a thoroughly dried state are exposed on barren land surfaces which are often found in arid and semi-arid regions. It can also occur locally on dry flood plains, tidal flats, lake beds and even glacial outwash plains. Thick deposits of wind-transported dust can accumulate under favourable conditions to form layers of loess—a porous, friable, yellowish sediment of mineral fragments mostly in the size range of silt, that is 0·06 to 0·004 mm in diameter. In China, loess deposits formed by deflation during the last glaciation, commonly reach 30 m, and in some places 90 m, in depth. Dust particles from the Rajasthan Desert in India are observed to be carried up to at least 5 km and to remain in suspension for several days, during which time they may be carried across Burma and Thailand into Cambodia. Similarly, dust from North Africa and Arabia is carried eastward along the coast of Iran.

Fluvial erosion and deposition are responsible for modelling the landscape in most climatic regions. Indeed, it is argued by some geomorphologists (i.e. Chorley, Leopold and Langbein) that the flow of water over the terrain can be viewed as a vast system which rapidly adjusts to the equilibrium condition of least work or optimum efficiency. This stability is characterized by a continual tendency for the land to minimize the energy expended by water running over its surface. Thus the land will continue to degrade gradually, reducing the available relief, while maintaining similar landforms, and this is reflected in the basic simplicity of landforms observed. Characteristic slope profiles are formed by different transportational processes. Slopes where the characteristic erosion process is soil creep or mass movement are generally convex upwards in profile, while wash-controlled slopes tend to be almost straight or concaved upwards. As the importance of concentrated flow and gullying increases on the slope so also does the degree of concavity, until the extreme case of rivers is attained where the longitudinal profile is very concaved.

Many geomorphologists have divided fully-developed slopes into four basic units, which are shown in Figure 10.4. The waxing slope is the term used for the convexed crest of the slope and it is generally believed that it is formed by soil creep and mass-wastage, that is by the movement of rock fragments and soil downslope solely under the influence of gravity. As the debris moves downhill it protects the underlying rock from further weathering and erosion, but the exposed rock left at the upper part of the slope is subject to weathering and so supplies further debris which moves down the slope. Since the top of the waxing slope is weathering at the fastest rate, the convex shape of the slope is retained. The free face is the name given to the outcrop of base rock, which is subject to rapid weathering, below the waxing slope. As the weathering products do not accumulate, the free face is where the whole complex is being most actively worn back. Material removed from the free face comes to rest on the debris slope, which may consist of an extensive scree slope or just a thin debris mantle. Weathering and movement continue on the debris slope which may show a reduction in particle size from top to bottom. The pediment is the basal element of the slope and in arid climates often consists of hard, fresh rock, with slight weathering, but in more humid climates it may be deeply weathered. Sometimes there is a distinct break of slope at the

Figure 10.4 Elements of a fully-developed hillslope. (After L.C. King, (1957). 'The uniformitarian nature of hillslopes'. *Trans. Edinburgh Geological Society.* 17, p. 81).

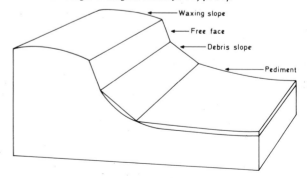

upper end of the pediment, and this is probably caused by intense weathering, together with the rapid removal of debris, at the base of the debris slope.

Slopes similar to that shown in Figure 10.4 are often observed in arid and semi-arid climates, but in more humid climates one type of erosion process may dominate over the whole slope. For example, in parts of northern Papua, straight slopes with angles between 35° and 40° predominate in the mountain areas. This indicates that erosion acts uniformly over the whole of the slope, and that it shows little tendency to diminish near the crests. Weathering on these slopes is balanced by erosion, and the material eroded is transported from the area by the rivers as fast as it is fed into them. The slopes themselves appear to be dominated by slopewash and the mass movement of soil. This combination of uniform slope erosion and the complete removal of eroded material by rivers results in 'ridge-and-ravine' type of landscape with parallel slope retreats causing narrow ridge crests.

Landforms of the Malay Peninsula

Landforms characteristic of the Malay Peninsula have been described by both Eyles (1969, 1971) and Swan (1970a and b). Eyles has made extensive use of various morphometric indices such as basin area, basin relief, average basin slope, drainage density and hypsometric integral. The latter is the ratio of the volume of earth material contained between the ground surface, the bottom and sides of the figure of reference (drainage basin) to the volume of the entire reference solid. After a careful statistical investigation, Eyles found that fourth order basins could be divided into six mutually exclusive groups which possessed reasonable internal uniformity. A fourth order basin is the catchment area of a fourth order stream segment, where the smallest or 'fingertip' channels in a drainage network are termed first order segments, second order segments being formed by joining any two first order segments, and so on for higher orders. The physical characteristics of each group are described by the group names which are: high mountains, low mountains, deeply dissected low mountains, isolated steep high hills, isolated steep hills, and low convex hills.

Eyles found that basins belonging to a particular group tend to occupy contiguous areas and therefore

it was possible to produce a map of drainage basin types. Such an analysis shows that the first three mountain groups only cover 32 per cent of the mapped area, thus emphasizing the essentially lowland nature of the peninsula. There is also an absence of extensive foothill zones, since the granitic mountain ranges commonly rise abruptly from the lowlands. Erosion in lowland granite areas usually produces low hills with unusually pronounced summit convexity separated by flat-floored swampy stream courses. Large areas of lowland granite are included in the low convex hills group, particularly in southern Johore. Comment was made in Chapter 2 on the tilting of the Malay Peninsula in late Cretaceous or early Tertiary times resulting in the extensive erosion and youthful landscapes of the granitic mountains of the north. The two groups containing isolated steep hills are associated with arenaceous sedimentary rocks or areas containing tower karst.

According to Eyles the 325 m contour roughly divides the Malay Peninsula into highland and lowland zones. He also considers that there is a fundamental difference between the mathematical forms of the longitudinal profiles of the streams in highland and lowland areas. Streams with a longitudinal profile of a form described by a quadratic equation are located mainly above about 325 m, particularly in the Main Range, the Bintang Range, and the Pahang–Trengganu border mountains. In contrast the longitudinal profiles of lowland streams are best represented by mathematical equations of the log-polynomial type.

Quadratic profiles are characteristic of mountainous areas where stream courses are steep, where nickpoints of a variety of origins are common, the bedload is coarse, and where sea-level exerts little or no control over the evolution of the profile. These youthful features are largely a result of the uplift of the northern Malay Peninsula at the end of the Cretaceous.

Longitudinal stream profiles represented by log-polynomial curves may be divided into three sections: (a) a very gently sloping lower portion; (b) an often well-developed concave nickpoint; and (c) a steep headwater reach. The lower section is formed on alluvium and reflects the well-developed flood plains which are so typical of the lands bordering the South China Sea, with their flat swampy courses usually undergoing aggradation. Some 15 000 years ago, sea-levels were between 100 m and 150 m below present levels and the rivers eroded their valleys to conform with this level. Sea-levels have been rising for the last 10 000 years, and over this period streams have had to continually adjust their valleys to the changing conditions, hence the aggradation. The nickpoint represents a discontinuity between the gently sloping lower reach, under the influence of rising sea-levels, and a steep irregular headwater reach, cut into highly resistant ridges and developing independently of the varying sea-levels. Eyles's analysis shows that flood plain development and the influence of rising post-glacial sea-levels have penetrated into the heart of the peninsula, the only areas excluded being the high mountains.

Swan (1970a and b) has studied landforms in Johore in some detail. This part of the Malay Peninsula did not undergo uplift at the end of the Cretaceous and since it has probably been under a humid equatorial climate since the mid-Tertiary, the only important changes are those resulting from eustatic changes in sea-level.

According to Swan a three-storey landscape exists in Johore. The top storey consists of steep hills, ridges and mountains, characterized by retreating slopes, the middle storey of dissected lowlands, and the bottom of low-lying tracts of late Pleistocene and recent deposits.

Mountainous terrain with a relative relief exceeding about 65 m occupies about a fifth of Johore and is associated with steep slopes with a mean gradient of about 21°. As should be expected, the lithology of the rocks has a significant influence upon relative relief. Arenaceous sedimentaries, acid volcanics and granites give rise to the highest relative relief, while marine alluvium yields the lowest. Similarly, arenaceous sedimentaries and acid volcanics are associated with the steepest slopes, marine alluvium with the gentlest, and the granites with intermediate slopes.

Terrain with a relative relief below 65 m occupies 45 per cent of Johore. Slopes are undulating to hilly, with a mean of just over 7°. The low-level terrain with an absolute altitude below 65 m is greatly dissected by streams and generally consists of a sea of low hills and rises. Average slope form is convexo-concave, though almost completely convex slopes are not rare. Indeed, many volcanic and arenaceous sedimentary rocks have slopes which are convex in shape since they do not readily give rise to weathered material which would accumulate at the foot of the slope. Granites, and argillaceous sedimentaries in particular, have large areas with concave slopes because material from these rocks is more readily eroded and deposited to form concave elements.

Where altitudes are over 65 m and in particular where they are over 130 m, the slopes become relatively uniform and straight. It is observed that there is an inverse relationship between the length of slope elements and the amount of slope curvature, thus lengthening the slope element reduces its curvature. Since in these areas of high relief slope angles are similar, it appears that general slope retreat is taking place producing a landscape with hills rising suddenly from flat lowlands. A similar type of landscape in northern Papua was described in the previous section.

11 Landscape Ecology and Tropical Ecosystems

ECOSYSTEMS which were mentioned in Chapter 1, are basic functional units of nature comprising both living organisms and their non-living environment. They can be considered as a series of components, such as living organisms, organic debris, water, rock minerals, nutrients and atmospheric gases, which are linked by food webs and flows of energy, water and nutrients. The living and non-living components interact among themselves and with each other and so maintain and develop the system.

Figure 11.1 illustrates some of the components of a forest ecosystem. The general vegetation depends on the conditions of the regional climate, the soil and the groundwater, but vegetation also affects soil formation, soil climate and the groundwater table. Both vegetation and soil are shown with a layered structure, which is the result of interactions within the system. Shading by the tree crowns reduces the light intensity below, producing local micro-climates each with its own fauna and micro-flora. Similarly, the movement of water and material in the soil produces the observed layers. Since the vegetation and soil both belong to the same ecosystem, any change in one is bound to cause a change in the other, and therefore soils and vegetation cannot be considered in isolation from each other.

Ecosystems are open systems connected to the surrounding environment by a number of inputs and outputs. These take the form of radiant energy, water, gases, inorganic chemicals and organic substances, and they are transported through the operation of energy, water and nutrient cycles. Changes in the nature of the inputs cause changes in the structure of the ecosystem as it adapts itself to the new conditions. So a change in regional climate could cause a forest ecosystem to change to a grassland ecosystem, and similarly forest clearance by man could disturb and change the ecosystem. When changing conditions cause alterations in the vegetation these will be followed by changes in the soil, since both are parts of the same system. If the basic inputs and outputs remain constant over long periods of time, and the system is not disturbed by man, it assumes a steady-state condition with a stable plant community, which in this case is known as the climax vegetation. So climax vegetation can be defined as a plant community which, given undisturbed conditions, is in stable equilibrium with climate and soil. Most ecosystems tend after an initial disturbance to return to a steady-state condition and the vegetation communities continually change until the climax vegetation is achieved.

Green plants are the only organisms capable of converting radiant energy from the sun into chemical energy through photosynthesis, and this represents the main energy input which drives the ecosystem. Energy was discussed in Section B, where the energy and radiation balances of plant communities were considered. The greater the energy input into an ecosystem, the more rapidly should the various flows, exchanges and interactions operate. This can be illustrated by a consideration of net primary production, which is the annual production of leaves, and organs of grass, as well as of wood and roots. If well-watered areas are considered, since lack of water limits the operation of an ecosystem, it is observed that net primary production is highest in the tropical and subtropical forests, and falls poleward through the broadleaved and coniferous forests of the temperate zone to reach very low values in the Arctic tundras. Since annual solar radiation increases towards the equator, this suggests a relationship between energy input and plant growth.

Matter and energy are stored in dead organic remnants such as forest litter, and the ratio of the litterfall of green leaves, old wood, etc., to the litter on the

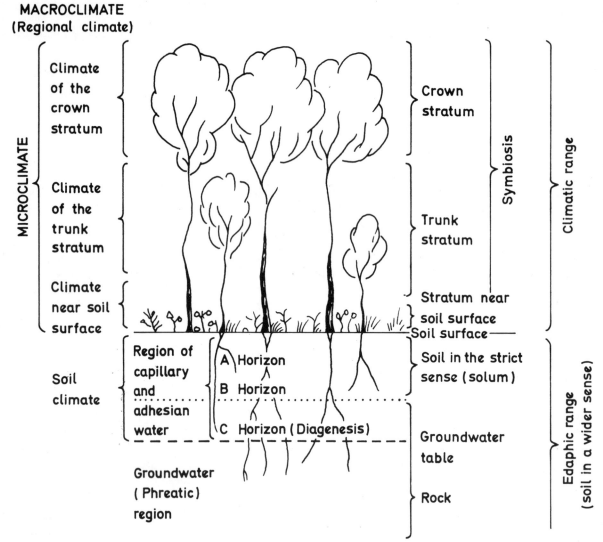

Figure 11.1 Schematic representation of the components of a forest ecosystem. (After Troll, 1963.)

ground is another measure of the rate at which flows and interactions are taking place in an ecosystem. The highest ratios are found in the dwarf-shrub tundra, followed in decreasing order by the coniferous forests, broad-leaved forests, and then the sub-tropical forests. In the tropical rain forests which have very high exchange rates, associated with a high energy input, the ratio is almost negligible.

Water is essential for the operation of an ecosystem. It acts as a coollant for plant leaves, and it is also required as a transporting mechanism and a medium for many of the exchanges and interactions

within the ecosystem. For energy to flow through a system there has to be an energy source and an energy sink. The major energy source for the ecosystem is solar radiation, while the major energy sinks are infrared radiation and latent heat loss via evapotranspiration. Detailed energy balances of plant leaves were described in Chapter 4, while the solvent action of water was described in the previous chapter, where it became clear that water is involved in many naturally-occurring chemical reactions.

Marked accesses or deficits of water severely limit the operation of an ecosystem, and this can be investigated by considering the ratio of net radiation available to evaporate water-vapour from a wet surface to the heat required to evaporate the mean annual precipitation. This ratio has values of up to 0·33 for the tundra, 0·33 to 1·0 for forest zones, from 1 to 2 for steppes, and above 2 for semi-deserts and deserts. Evidence exists that there is an optimum value of this ratio near 1·0 at which, for a given value of solar radiation input, vegetation productivity is at a maximum. With ratios near unity there is just enough net radiation to evaporate the mean annual precipitation, so transpiration is not curtailed by lack of moisture as it would be for ratios less than unity, nor is the soil waterlogged and poorly-aerated as it would be if the ratio were greater than unity. Thus the activity of ecosystems should vary with the solar energy input and with the ratio annual net radiation/annual precipitation.

Evapotranspiration and run-off are the two major water outputs from the ecosystem and may be redefined from this view-point. Thus potential evapotranspiration is the water loss from an ecosystem in a steady-state condition and with a satisfactory water supply, required to maintain, together with the heat loss, the energy balance of the system. Evapotranspiration is thus essential to the operation of an ecosystem. In contrast, run-off represents water which is completely surplus to the operation of the ecosystem, suggesting an excess of rainfall. Under dry weather conditions, run-off will cease but the ecosystem will continue to operate for long periods using water stored in the soil. Only when the soil moisture is exhausted does the water supply to the ecosystem become limited, whereupon plants over-heat, wilt and die. If unusually dry conditions continue for a long period, as in a drought, the whole ecosystem may be fundamentally damaged and may take several years to recover when the climate returns to normal.

Nutrients are found in four basic compartments within the ecosystem, and each compartment is linked to the others by an array of natural processes. These compartments may be summarized as follows:

(a) The organic compartment. This consists of the living organisms and their debris.

(b) The available-nutrient compartment. This is composed of nutrients held on the surface of soil particles or in solution in soil water.

(c) The unavailable-nutrient compartment. This consists of soil and rocks containing nutrients in chemical forms which are temporarily unavailable to living organisms.

(d) The atmospheric compartment. This is made up of gases which can be found not only in the atmosphere but also in the soil.

Nutrient flow between the various compartments is driven by solar energy, but it is also closely connected with the water cycle described in Section D, since water carries nutrients through the ecosystem in solution. Nutrient exchange takes place between all the compartments but a particularly well-defined cycle exists between the organic and available nutrient stores. Thus forests may attain high levels of organic production on soils regarded as too poor for agriculture, since large amounts of easily soluble nutrients are involved in a biological circulation between the forest soil and its plant cover.

Many kinds of organic matter (such as leaves and twigs) are produced in ecosystems by photosynthesis in chlorophyll-bearing plants, and this organic matter can be eaten by secondary organisms (such as birds and animals) which depend on plants for food. The organic matter produced in the plant canopies eventually falls to the soil surface as plant and animal litter, and results in the build-up of a nutrient-rich layer of dead organic matter covering the soil. This superficial litter layer slowly decomposes and provides nutrients to the soil, thus providing the major flow path of nutrients from the organic to the available-nutrient compartment. Nutrients may also be leached from plants directly by precipitation, thus providing a second means of transport to the soil. A third method of transfer is by the annual death and decay of fine roots within the soil. The cycle is completed by the return of nutrients to plants via the root system.

Material from the unavailable-nutrient compartment is transferred to the available-nutrient stock by the process of weathering, which was described in the previous chapter. Nutrients can be added to an ecosystem by precipitation, and also by dust captured in tree canopies. The nutrients in precipitation have various sources, one of which is the cyclic salts of the oceans mentioned in the previous chapter. Electrical discharges in thunderstorms form nitrates which can be washed out of the atmosphere by precipitation, while other sources of nutrients are volcanoes and forest fires. Bacteria living in the root systems of certain plants also have the ability to assimilate gaseous nitrogen which eventually becomes available to plants.

Nutrients are lost from ecosystems mainly in drainage water, which eventually forms river-flow. Forest fires are another important cause of nutrient loss from forest ecosystems, since nutrients such as nitrogen are released into the atmosphere and mineral elements may be washed out of the ash covering the soil. Minerals in river water can be in solution and in particulate form, and they may arise from cycling within the ecosystem or by direct erosion which is normally small in undisturbed systems. The mineral content of rivers has already been discussed in connection with erosion, and it is now seen that the erosion of natural landscapes is just a part of the operation of a complex ecosystem. Since the operation of the land ecosystem determines the mineral content of rivers, it will also partly determine the nature of the sediments being deposited in nearby seas.

Nutrient cycling in tropical rain forest in the Ulu Gombak Forest Reserve, Peninsular Malaysia, has been studied by Kenworthy (1970). He found that patterns of nutrient cycling are different for each element, and this may be illustrated by a consideration of potassium, calcium and magnesium. In the case of potassium, the input through precipitation and the output through solution in stream water are both of the order of 12 kg/ha/annum. Within the ecosystem the cycling of potassium is considerably greater, for the concentration in rain-water passing through the forest canopy to the soil increases eight times, and all of this is then taken up again by the vegetation. Calcium behaves in a similar manner, except that at no point in the cycle is the concentration greater than three times the initial input in rain-water. In the case of magnesium most of the internal increase comes from the leaching of litter. Kenworthy also noted that the nutrient cycle followed to a great extent the patterns imposed by the hydrological cycle. Thus the release of nutrients by leaching from a canopy only takes place when the interception capacity of the canopy is exceeded. Similarly leaching of surface litter and movement of soil water into streams will only take place on a significant scale after extensive rainfall.

So far in this discussion man's influence on ecosystems has been assumed to be small, the landscapes discussed being completely natural. In many areas man largely controls the ecosystem through crop cultivation, and even in what appears to be wild areas his influence can be extensive. In cultivated areas the whole structure of the ecosystem may be changed, resulting in changes in the soil, the fauna and the outputs. The process of harvesting crops or felling trees, removes both energy and nutrients from the ecosystem, and this in particular disturbs the nutrient cycle, which can be further modified by the addition of fertilizers. When ecosystems are well managed, by good agricultural technique, they remain stable, but if badly managed, they may become unstable and widespread changes may occur such as extensive soil erosion. For instance, if large amounts of nutrients are stored in both the vegetation and in the dead organic litter, as in tropical rain forest, removal of the trees may completely destroy the nutrient cycle and leave a very infertile soil which erodes rapidly.

As climate changes, so does the natural vegetation, thus creating new ecosystems with new environments and new animals. Indeed, fauna are often limited to very restricted environments and only continue to exist while their environments remain unchanged. Extensive areas with a similar environment allow birds and animals to migrate, but if the environment becomes restricted, so often does the geographical range of the particular bird or animal. For example, at present Ceylon is separated from India by a narrow sea strait and by a zone of arid climate in both north Ceylon and south India. Various forest animals live in Ceylon which are very similar to those which are found in India, yet they could not have reached Ceylon under the present environmental conditions. Instead, they came when the sea-level was lower and the climate more humid, creating a continuous strip of forest from India to Ceylon. Even so, not all Indian forest animals managed to migrate along the forest bridge, so the leopard reached Ceylon but not the tiger, the otter but not the weasel, the Indian sloth bear but not the hyena.

Animals sometimes adapt themselves to changes in environmental conditions. For example, a small owl is found in dry deciduous or thorn scrub forest in both south India and Ceylon. It is well adapted to life in this type of environment, and it is not found in tropical rain forest. A related Ceylonese rain forest owl does also exist which is almost identical with the owl found in the thorn scrub except that it has much darker plumage. It appears that during a period when the tropical rain forest was widespread, the forest owl migrated from India to Ceylon. The climate then became drier, the rain forest retreated, and there was only a small isolated population of forest owls in Ceylon. In India the owls gradually adapted themselves to the new arid environment, and when they had done this, they invaded the arid regions of Ceylon, thus giving Ceylon two versions of the same bird, each living in its own environment.

Soils

Most rocks are covered by a layer of weathered material, which is extremely shallow in the Arctic but may reach many tens of metres in depth in the tropics. The term regolith is used to describe both residual and transported loose or soft material overlying solid bed rock. Because of the operation of energy, water and nutrient cycles within the ecosystem, the surface layers of the regolith are further altered to form a material called soil. So soil may be defined as a material consisting of both mineral and organic matter covering the surface of the regolith.

Soil formation may be regarded as the result of the various energy, water and nutrient cycles operating within the ecosystem. Thus the soil surface gains water by direct precipitation, by secondary drip from plants or by flow over or through the soil. It also gains gases from the atmosphere, organic matter from decaying litter, energy from the sun and mineral matter in the form of dust deposits. It loses water by evaporation, transpiration and seepage to lower layers, gases to the atmosphere, energy by radiation or conduction to lower layers, and material by erosion. The balance of these various processes partly determines the type of soil formed. Thus if precipitation exceeds evapotranspiration, water will move down into the soil, removing material in solution and also washing down small mineral particles, processes which are known as leaching and eluviation respectively. The mineral particles may accumulate in lower layers of the soil, a process known as illuviation, while material in solution may also be deposited by chemical precipitation. Similarly, if evapotranspiration exceeds precipitation, water will move towards the soil surface and deposit minerals in the upper layers of the soil. The operation of processes of this type produces a series of horizontal layers in the soil known as soil layers or soil horizons. The assemblage of soil horizons is called the soil profile.

Decaying organic matter is usually found on the soil surface due to the accumulation of plant and animal litter. Organic matter which has been partially decomposed and has become part of the soil is usually termed humus, and is mixed into the soil from the surface litter layer and from dead plant roots. There are numerous grades of humus, which tend to vary according to the nature of the associated plant community. Some plant communities extract large quantities of mineral nutrients from the soil, and these are incorporated in their leaves, stem and roots. When such plant organs die, the nutrients tend to be retained by the resulting humus which has a neutral or only mildly acid reaction and is known as mull. Other plant communities extract little in the way of nutrients from the soil and the resulting humus is poor in minerals and strongly acid in reaction, being referred to as mor.

In the soil, humus enters into associations with tiny clay particles to form what is usually referred to as the 'clay-humus complex'. This is chemically very active since it is capable of retaining mineral elements in the soil in an exchangeable and readily available form. Without it, one shower of rain would be sufficient to wash most of the highly soluble nutrients away. It is largely the presence of the clay-humus complex which turns an infertile regolith into a fertile soil. Ions of potassium, calcium, magnesium and other metals, and also the ammonium ion, all of which are needed by plants, are held by the clay-humus complex, thus increasing the soil fertility. Now rain-water contains dilute carbonic acid, and the hydrogen ions in this replace the metallic ions in the clay-humus, the metallic ions being leached downwards. When the clay-humus particles lose their nutrient ions because of percolating rain-water increased acidity must occur because hydrogen ions are substituted. A soil whose clay-humus complex is almost saturated with hydrogen, with no other acids present, has a pH of approximately 4, but if it contains free organic acids derived from humus, pH values may reach 3·2 or less. In contrast a soil which is base-saturated may have a pH of anything between 7 and 8, or even higher if it contains free sodium carbonate. Most normal natural soils have pH values of between 4 and 7. pH may thus be regarded as a very rough guide to the soil fertility, which can only be determined accurately by careful soil analysis.

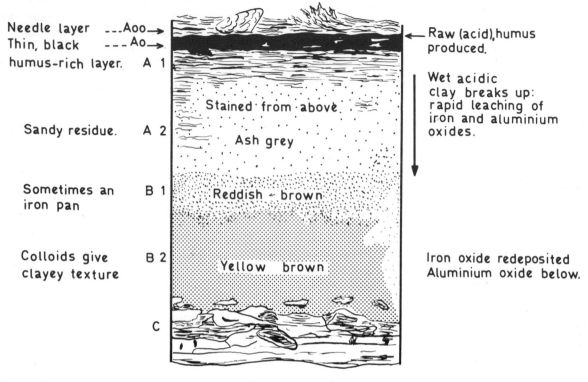

Needle layer ---Aoo→
Thin, black --- Ao→
humus-rich layer.   A 1

Sandy residue.   A 2

Sometimes an   B 1
iron pan

Colloids give   B 2
clayey texture

C

Stained from above

Ash grey

Reddish - brown

Yellow brown

← Raw (acid),humus
produced.

Wet acidic
clay breaks up:
rapid leaching of
iron and aluminium
oxides.

Iron oxide redeposited
Aluminium oxide below.

Figure 11.2 Vertical section through a podzol. (After Money, 1972).

SOIL HORIZONS

Characteristic letters are sometimes used to describe the soil horizons which make up the soil profile, though unfortunately the nomenclature varies somewhat between nations, organizations and even individual research workers. Often the *A* horizon is regarded as the layer from which material is washed out (eluviated) to be deposited (illuviated) in the *B* horizon. The *C* horizon is the parent material which may be weathered rock, in which case the parent bed rock forms the *D* horizon.

Soil horizons are well illustrated by a study of a soil type known as a podzol (Figure 11.2). Typically these are found in coniferous forests where there is an almost continual downward movement of water because of a rainfall excess or a very sandy pervious soil. Coniferous trees extract few mineral nutrients from the soil and as a result their litter decays into a very acid mor. Because of the leaching due to the downward movement of water and the presence of very acid humus, the whole soil becomes very acid, and under these conditions organic acids are able to re-

move oxides of iron and aluminium from the surface layers leaving silica in the form of sand, and forming an upper eluviated horizon. The material removed from the upper horizons is re-deposited in the lower part of the soil profile forming a set of lower illuviated horizons. Normally two distinct illuviated horizons may be recognized, the upper one being reddish-brown or black in colour while the lower may be any colour ranging from a dark, reddish-brown to a pale yellow brown. These horizons arise from the re-deposition of humic material in complex chemical combinations with complex oxides (sesquioxides) of both iron and aluminium. The various oxides may have a cementing effect in the upper illuviated horizon, and sometimes the iron sesquioxides form hard concretions of ferruginous material known as hard pan.

Various parts of the podzol profile may be given characteristic labels. The uppermost part of the soil profile consists of a loose mat of undecayed pine needles known as the A_{00} horizon, which merges into the A_0 horizon formed of decomposed organic material. There is often a sharp discontinuity between the A_0 and A_1 horizons, the latter being almost entirely sand which has been stained with organic material derived from above or decaying plant roots. The A_1 horizon merges into the A_2 horizon which is formed almost completely of bleached sand. The junction between the A_2 and B_1 horizons can be very sharp, the latter being shallow and merging with various degrees of sharpness into the B_2 horizon. The B_1 horizon sometimes contains layers of iron pan. At the bottom of the profile is the C horizon which consists of partially weathered parent-material.

SOILS OF THE HUMID TROPICS

Chemical weathering of the primary minerals of rocks is the same under both temperate and tropical conditions, that is to say it results from the actions of water, carbonic acid, and organic acids acting over a period of time. Nevertheless, the tropical environment is different from that found in the temperate latitudes in two important aspects—the high temperatures and humidities, and the stability of the climate over long periods of time. A rise of 10°C in temperature causes an approximate doubling of the rate of chemical reactions, so weathering processes in the tropics are approximately three to four times as rapid as those in temperate latitudes. Climates in the humid tropics were relatively stable during the Quaternary glaciations, hence many soils have been under a uniform vegetation for many millions of years and probably even from the mid-Tertiary. In contrast, many temperate latitude soils are only 10 000 years old, and some are considerably younger.

The main weathering products, except under extreme leaching, are clay minerals derived mainly from felspars, and iron oxides in varying degrees of hydration derived in the main from the ferro-magnesian minerals. Quartz, together with other virtually indestructible minerals, is generally unaffected by chemical weathering. Many clay minerals become unstable under conditions of heavy leaching and high temperatures over a considerable period of time, and decompose. Kaolinite appears to be the most stable clay mineral under these conditions and it is found in many tropical soils. Normally all silica other than that forming part of clay minerals and sands is leached out of the upper layers of the soil and either re-deposited in the weathered material below or completely lost in the rivers. Hydrated oxides of iron, aluminium and manganese remain in the upper layers of the soil after the silica is removed. The iron sesquioxides may be in the form of limonite, which is yellow in colour, or red haematite. Under suitable conditions both the hydrated oxides of iron and aluminium become mobile in the soil profile. In particular, there are several mechanisms whereby iron oxide can become mobile and be concentrated within the soil profile. It tends to be deposited at the surface of a water-table or within the zone of fluctuating saturation above such a water-table. In this zone, iron oxide leached downwards meets that carried upwards with the rising water-table, and very favourable conditions are created for their deposition.

Observations reveal that the normal processes of soil formation within the tropics produce soils which have the same characteristics as those found in temperate regions. Thus podzols are found in the humid tropics. Soils are also found in the humid tropics which are not similar to those observed at the present time in temperate latitudes, though similar soils may have existed in the temperate latitudes during the Tertiary.

Podzols. In South-East Asia these soils are observed on level or slightly undulating terrain in regions where the annual rainfall exceeds 2 000 mm. They are most extensive on old coastal terraces covered with sandy materials, but also occur on sandstones, quartzites and acid volcanic rocks. Podzols are light in texture and under natural forest they develop an organic surface layer of matted mor. Their sub-surface horizon is light grey and strongly bleached, grading into a dark brown to reddish illuvial horizon composed of organic matter and iron oxides, or of organic matter alone. The depth of podzols varies widely, and those formed on level ground can have bleached horizons over 100 cm thick and illuvial horizons extending to 250 cm below the surface. The soils are very acid, with average pH values of 4—5 or less in the surface horizons.

Red-yellow Podzolic and Red-yellow Latosols. These soils form an acidic to moderately basic parent materials, and on residuals of sedimentary, igneous or metamorphic rocks. They are rarely found in areas with an annual rainfall of less than 1 500 mm or in areas with a marked dry season. The control exerted by topography appears to be slight, as they form in both flat and extremely mountainous terrains. They are found under natural vegetation ranging from tropical

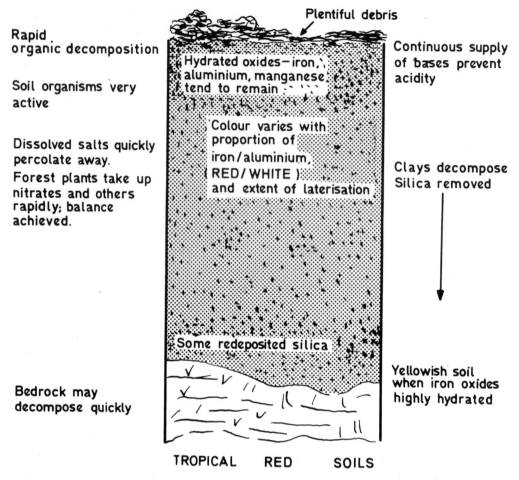

Plentiful debris

Rapid organic decomposition

Hydrated oxides—iron, aluminium, manganese tend to remain

Continuous supply of bases prevent acidity

Soil organisms very active

Colour varies with proportion of iron/aluminium, (RED/WHITE) and extent of laterisation

Dissolved salts quickly percolate away.
Forest plants take up nitrates and others rapidly; balance achieved.

Clays decompose
Silica removed

Some redeposited silica

Bedrock may decompose quickly

Yellowish soil when iron oxides highly hydrated

TROPICAL   RED   SOILS

Figure 11.3 Vertical section through a tropical red soil.
(After Money, 1972.)

forest to short grass savanna. This is the dominant soil type in the wetter parts of South-East Asia wherever non-basic parent materials outcrop, and it covers considerable areas in regions such as Peninsular Malaysia.

Normal soils of this type (Figure 11.3) under natural vegetation have an A horizon which is distinctly differentiated into a humiferous A_1 layer and a somewhat paler-coloured leached A_2 layer. At low altitudes the A_1 layer is usually weakly developed, but at higher altitudes it may be 25 cm thick and extremely humic, especially on level ground. The B horizon varies in colour from red to yellow, and in thickness from a few centimetres on sloping ground to over a metre in level areas. The profiles often contain concretions or continuous layers of a substance called laterite. The pH values of these soils are in the range of 4·5 to 5·5.

As a group these soils are very poor in nutrients and crop yields are low without skilled agricultural management. They are also highly susceptible to erosion because of their low permeability and the sharp transition between the rather open surface horizon and the heavy sub-soils. Even on slight slopes, intense rainfall can remove the surface layers, and especially if the protecting vegetation has previously been removed.

Laterites, Plinthites and Ferrallites. The word 'laterite' was originally suggested as a name for a highly ferruginous deposit first observed in the Malabar region of India. When fresh it is soft and very easily cut, but on exposure to the air soon becomes very hard and resistant. If it is cut into bricks and then allowed to harden, it forms an ideal building material. The name is derived from the Latin—*later*—meaning a

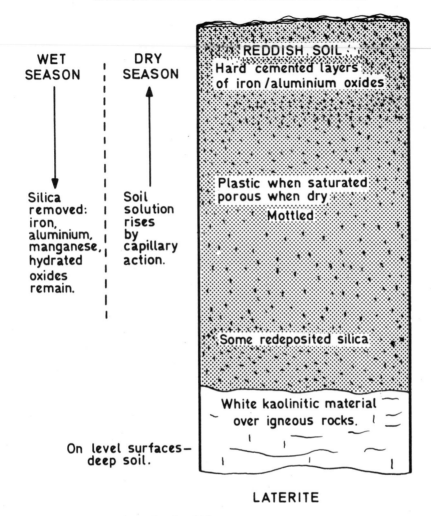

WET SEASON | DRY SEASON

Silica removed: iron, aluminium, manganese, hydrated oxides remain.

Soil solution rises by capillary action.

REDDISH SOIL
Hard cemented layers of iron/aluminium oxides

Plastic when saturated porous when dry
Mottled

Some redeposited silica

White kaolinitic material over igneous rocks.

On level surfaces— deep soil.

LATERITE

Figure 11.4 Vertical section through a laterite. (After Money, 1972).

brick, and obviously relates to its use as a building material and not to its red colour. Unfortunately the name has been applied to many reddish soils many of which do not have the properties of the original laterite, hence the use of the terms plinthite or ferrallite to describe what was originally laterite.

Laterite (Figure 11.4) may be defined as a sesqui-oxide-rich material forming aggregates of varying size and colour, or continuous layers with reticulate colour patterns. It includes hard concretions, pans and a soft material which hardens irreversibly on exposure, and appears to be normally formed in soil horizons which are seasonally or temporarily saturated by a fluctuating water-table, iron oxides being precipitated at the upper boundary of the water-table. In the layer

over which the water-table fluctuates, 'active' or soft sheets of continuous laterite develop, usually with sharp boundaries and running roughly parallel to the surface. The laterite horizon is only exposed following recent soil movements such as those due to soil creep, sheet erosion or the root movements associated with falling trees. Concretionary laterite may develop as a suspended horizon above a less permeable layer in wet climatic regions. Locally, it is also observed to develop at or near springs where groundwater containing iron comes into contact with the atmosphere. Similarly, it may form around depressions in basalt plateaux where localized run-off or lateral drainage deposits iron oxides on the lower slopes.

Actively-forming laterite horizons are usually at

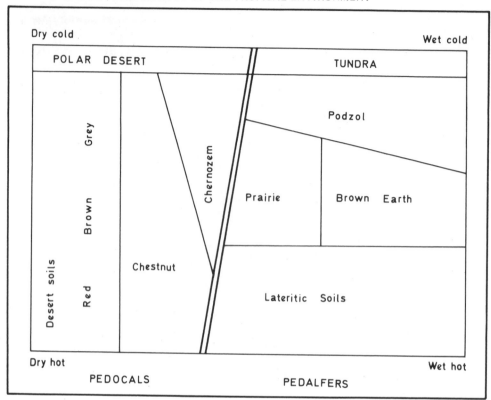

Figure 11.5 Climatic relationships of the zonal soils. (After Ollier, 1969).

some depth below the surface, but subsequent erosion may expose them at the surface where they will harden in the atmosphere. These extensive layers of hard laterite concretions are very resistant to erosion and may eventually be left as remnants of the lower surface capping low hills.

Soils in which laterite occurs are not necessarily poor soils, and laterite does not affect the agricultural value of a soil, unless it appears at a shallow depth or as a continuous hard surface layer, when there may be difficulties in root penetration.

WORLD DISTRIBUTION

Soils may be divided into Zonal, Intrazonal and Azonal soils. Zonal soils reflect the regional climate and vegetation to a large degree, and are formed on

Figure 11.6 Schematic representation of the profile relationships of zonal soils in a traverse from pole to equator. (After Ollier, 1969).

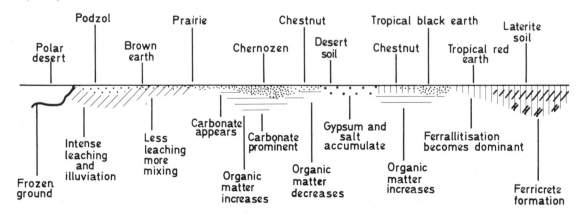

well-drained sites on non-extreme parent material. In contrast, intrazonal soils are well-developed soils formed where some local soil-forming factor is dominant, while azonal soils have poorly-developed profiles because some factor such as parent-material or lack of time has inhibited soil formation.

The climatic relationships of zonal soils are shown in Figure 11.5, while Figure 11.6 indicates the profile relationships of zonal soils in a traverse from pole to equator. Zonal soils may broadly be divided into pedalfers and pedocals. A pedalfer is a soil in which clay particles are broken down by leaching or laterization, and which contains no calcium carbonate. If evaporation exceeds precipitation for a long period of time, various salts such as calcium carbonate and calcium sulphate may be deposited in the surface layers of the soil, giving it an alkaline reaction. Such soils are known as pedocals.

Some of the zonal soil types have already been described, i.e. podzols and lateritic soils. Brown Earths typically occur under deciduous forest in areas to the south of the main podzol coniferous forest zone. The soils exhibit leaching and are usually acid, but all these features are less intense than in the podzols. Prairie soils merge into the Brown Earths, but occur in the drier areas beyond the range of natural forest. They have a three-part profile:

A. dark, humus well-distributed, only slightly acid;

B. transition horizon, often brown;

C. parent material.

Chernozems develop under steppe, prairie or semi-desert in semi-arid climates. Upward movement of water under drying conditions causes calcium carbonate to accumulate as nodules, but occasional showers leach out sodium and potassium salts. The soil is basic and the clay content does not break up, a marked distinction from the podzols. Forming under grasses, the A_1 horizon is dark, nearly homogeneous and about a metre thick. Beneath this a brownish A_2 layer with calcium carbonate nodules merges into a poor B horizon which may contain calcium carbonate concretions, for there is very little leaching. Below is the C horizon formed of weathered parent material. Chestnut soils are found on the arid side of the chernozem soils under a thin grass cover. They differ from the chernozem soils in having the carbonates closer to the surface and containing less organic matter. The Brown and Grey soils of the semi-deserts may be re-garded as extreme forms of chestnut soils where carbonates come even closer to the surface, and the organic content is lower. Tropical Black Earths are found in the tropical grasslands with seasonal droughts, and they resemble chernozems in that they contain calcium carbonate in the lower layers.

There are many types of intrazonal soils, which owe their formation to some dominant local soil-forming factor. Under waterlogged conditions the soil is deprived of oxygen and bacterial action is hindered resulting in the accumulation of partly decomposed plant remains, forming a peaty layer on the surface. Below the peat is a layer of bluish-grey clay with iron oxide mottles in yellow, red or brown, resting on a permanently wet zone coloured blue-grey and containing iron sulphide. Such a soil is known as a Gley soil. Other intrazonal soils include those containing excess sodium salts and others formed on peats and consisting almost entirely of organic matter.

DESERT SOILS

Tropical areas are often divided into two broad climatic classes—humid and arid or semi-arid. Typically the humid areas are covered by various types of forest, while the semi-arid and arid areas are covered by grassland or only very sparse vegetation.

Chemical weathering is probably the dominant form of weathering even in deserts, and it tends to produce large quantities of coarse particles. On slopes where erosion predominates, soils are thin or absent and this tends to be the normal condition in deserts. Only on stable surfaces such as those of low-angle alluvial fans and pediments do materials accumulate and soils form. Contrasts between soils on different parts of a slope are often more pronounced in dry areas than in humid ones, because soils can only form on shallow slopes. For instance, water flows off the high-angle slopes of desert inselbergs and accumulates in the coarse alluvium of the plain, where it leads locally to deeper weathering. Soils formed on steep slopes are also often younger than those formed on low-angle slopes, because of faster rates of erosion and deposition.

Grey earths or serozems are found in arid and semi-arid areas. These soils contain calcium carbonate and calcium sulphate in the profile, but with the greatest concentration within 30 cm or so of the surface. The surface of the soil is often covered by a layer of wind-scoured pebbles which protects the underlying soil from wind erosion.

Solonchaks, saline soils or white alkali soils are found mainly in depressions or near the base of slopes. In these soils the capillary rise of salt-rich water and its evaporation causes the precipitation of salts on the soil surface, so the *A* horizon is loose and puffy, and may have a salt crust.

Solonetzs or black alkali soils, have a dark colour caused by finely-dispersed organic colloids and develop when an improvement in drainage causes the leaching of soluble salts from a solonchak.

Many desert areas contain soils, the top profiles of which are cemented with either silica, sesquioxides (particularly iron) or carbonates, and these hard layers or duricrusts protect the softer material beneath from erosion. Because of their great hardness, duricrusts often outcrop on slopes and cap low hills. The typical duricrust in semi-arid Australia has associated with it four main horizons. The top horizon consists of a deep red ferruginous zone which forms a hard laterite crust. Below this is a mottle zone of white kaolinic material with red patches. These ferruginous mottlings decrease downwards until normally a pallid zone is reached, but sometimes this zone is absent. Beneath the pallid zone is the parent rock, but sometimes a layer of silicified material forms a basal horizon above the parent rock. Duricrusts formed from laterite vary between 1 m and 13 m in thickness, whereas silica duricrusts are slightly less thick.

Some duricrusts are probably very old, since they are sometimes observed to be folded, or to occur beneath considerably dissected ancient landscapes, or under fixed sands, or even lava flows.

Australian landscapes can be grouped into four main cyclic surfaces, namely a Gondwana Surface formed in a pre-Cretaceous cycle of erosion, an Australian Surface formed by erosion lasting from the Cretaceous to the Miocene, a multi-phase Lake Eyre Surface dating from late Tertiary time, and recent local surfaces. Duricrusts formed of hard laterites are found on the remains of the Australian Surface, and it seems that this laterite started to form in the late Cretaceous or early Tertiary under conditions of heavy seasonal rainfall. At this stage the laterite zone would have been buried in the lower layers of a soil, as is observed in actively-forming laterites today. Since its formation the climate has changed and become drier, and the surface layers of the original soil

have been eroded away leaving a hard layer of laterite as a duricrust. Laterite duricrusts are very widespread in the tropics and mostly seem to date from more humid periods in the past. These humid periods may have been the result of actual climatic changes such as those associated with the marked cooling at the end of the Tertiary or with the ice ages, or in the case of India and Australia from the rapid drift northwards of continents through tropical climatic zones. Once they have formed, duricrusts are very stable in semi-arid or arid climates, provided that extensive periods of frost are absent.

Vegetation and Ecosystems

Stable vegetation communities which are in equilibrium with climatic and soil conditions are referred to as the climatic climax vegetation, and they form part of an ecosystem in steady-state equilibrium. Since climate varies considerably over the surface of the earth, with many regions of restricted water or energy supply, several types of climatic climax vegetation are to be expected. Thus near the equator, with an adequate water and energy supply, tropical rain forest is the climatic climax vegetation, but near the sub-tropical deserts this is reduced by lack of water to various mixtures of grassland with occasional trees or semi-desert scrub. Similarly, lack of water limits the climatic climax vegetation to grassland over large areas of the middle latitude continental interiors. In the north temperate regions, plant growth is often limited by a lack of energy rather than by lack of water, since the incidence and length of hot and cold seasons are more important than that of wet and dry seasons as in the tropics. Thus climatic climax vegetation changes poleward from deciduous to coniferous forest and then to tundra vegetation. Since all these changes in climatic climax vegetation represent changes in the whole structure of the local ecosystem, they closely reflect the changes in zonal soil type which have just been described.

TROPICAL RAIN FOREST

The climatic climax vegetation of much of the lowland humid tropics consists of various types of tropical rain forest. Undisturbed primary forest in the lowlands of South-East Asia is composed of many thousands of species of trees, as well as shrubs, herbs and woody climbers. Large groups consisting of one species of tree are rare and often up to one hundred different species with a girth of more than 25 cm are found to the acre. The vertical structure of the forest can be divided into a number of layers. The upper or emergent storey is usually about 30 to 50 m high, though trees nearly 70 m in height are often present. This upper storey does not form a continuous canopy and usually consists of trees about half of which belong to the family Dipterocarpaceae. The second or main storey, which occupies a layer about 20 to 30 m from the ground, normally forms a continuous canopy except immediately below large emergent storey trees. It consists of young trees of the normal emergent storey species together with trees of various other families. The understorey or third tree layer consists of saplings of the upper two storeys together with trees which do not normally grow beyond this level. Comparative freedom of movement on the ground is possible because the shrub layer is usually open, since the low light intensities at ground level do not favour vigorous growth. Under optimum conditions tropical rain forest shows no seasonality of flowering or leaf-fall since both occur throughout the year.

Lowland forest in many equatorial areas has been subject to destruction by storms, by aborigines practising primitive shifting cultivation and by recent industrial and commercial development. In many areas the original vegetation has given way to a patchwork mosaic of secondary vegetation in all stages of development from open land, through low scrub, bamboo scrub, bamboo clumps interspersed with scattered trees, to closed forest.

Local conditions can produce distinctive vegetation complexes. The main difference between lowland dipterocarp and hill dipterocarp forest is a shift in the floristic composition of the upper and main storeys. Many of the common lowland forest species are still found in the hill forests, but less frequently, while many species occur which are absent in lowlands. Thus the most common large tree species in hill forest is *Shorea curtisii*. The vegetation of steep hill slopes is often poorly stocked in woody species. At elevations above about 1 400 m various northern tree types make their appearance among the more characteristic tropical ones, and oaks, maples, magnolias and rhododendrons are observed. On exposed ridges and summits above cloud level a type of mist or moss forest is found, where the forest is reduced to a single layer about 10 m high and the frequently gnarled and stunted trees are covered by liverworts, mosses and ferns. The soil in these forests is covered by accumulations of acid humus and peat.

Muddy shores, lagoons and the estuaries of tidal rivers are frequently covered by Mangrove Forests. These are simple in structure, 7 to 14 m in height with a comparatively even and unbroken top canopy and a very poorly developed or completely absent under-storey or shrub layer. The principal tree species are found nowhere outside the mangrove forests and are frequently characterized by special root formations such as stilt roots.

VEGETATION SUCCESSION

Over large areas of the earth's surface the climatic climax vegetation does not exist because it has been removed by man, usually through the action of cultivation. More locally, it may not exist because of special local conditions, such as extremely steep slopes or very wet soils in a river valley. Thus while tropical rain forest is the climatic climax vegetation of South-East Asia, large areas are not covered by forest because of the direct interference of man.

If there is no human interference, vegetation communities usually change and develop until they are in equilibrium with their environment. A good example of this is provided by the island of Krakatoa, Indonesia, which was involved in a gigantic volcanic explosion in 1883. Part of the island was removed and the rest covered by white-hot volcanic ash, completely destroying the soil and all life. A few months after the ash had cooled it became covered by a film of algal slime, since these primitive plants are able to live on a completely inorganic surface, provided that the climate is humid. Soon this pioneer community was replaced by a second stage community of herbaceous plants and grasses which rapidly formed a cover over the ground. After a further ten to twenty years the second stage community gradually gave way to a third stage community of shrubs and trees, which slowly developed into tropical rain forest. Each stage in the development of the plant community produced conditions leading onto the next stage until the whole ecosystem came into equilibrium. Tropical rain forest could not grow at once on the island because soil was lacking; only when weathering and the action of algae had produced some form of soil, could a sequence of events start which eventually led to tropical forest.

Vegetation communities must therefore be viewed as part of a dynamic ecosystem, which may or may not be in equilibrium with its environment. Concepts of natural or wild vegetation, with associated soils, usually imply ecosystems which are in a steady-state equilibrium. Most ecosystems have been disturbed by man and may not be in equilibrium with their environment, so the landscape is covered by a complex mosaic of ecosystems whose exact equilibrium status must be in doubt. Ecosystems which are unstable because of interference attempt to evolve towards a new stable condition, so interference with plant communities may lead to changes in soil type, stream run-off, erosion rates, etc., as adjustments take place. Ecosystems which are drastically altered may become extremely unstable for a short period of time before they start to develop towards a new equilibrium condition. Thus the complete clearance of tropical rain forest may lead to all the soil being removed by rapid run-off and erosion, leaving a completely sterile landscape. Left to itself this landscape will eventually return to tropical rain forest, as on Krakatoa, but the initial interference had extremely destructive results.

It is clear then that the natural environment operates through several extremely complex natural systems and that physical geography is very much concerned with the study of the operation of these systems. This is of particular importance since man needs to be able to control these systems and maintain them in a steady-state equilibrium if he is to survive in a civilized state on the planet Earth.

FURTHER READING FOR CHAPTERS 10 AND 11.

Birot, P. (1968). *The Cycle of Erosion in Different Climates* (Batsford, London).

Cooke, R.U. and Warren, A. (1973). *Geomorphology in Deserts* (Batsford, London).

Dudal, F. and Moormann, F.R. (1964). 'Major Soils of Southeast Asia'. *Journal of Tropical Geography*, 18, p. 54.

Eyles, R.J. (1969). 'Depth of Dissection of the West Malaysian Landscape'. *Journal of Tropical Geography*, 28, p. 23.

_____(1971). 'A Classification of West Malaysian Drainage Basins'. *Annals Association American Geographers*, 61, p. 460.

Eyre, S.R. (1963). *Vegetation and Soils, a World Picture* (Arnold, London).

Jennings, J.N. and Mabbut, J.A. (1967). *Landform Studies from Australia and New Guinea* (Cambridge University Press, London).

Kenworthy, J.B. (1970). 'Water and Nutrient Cycling in a Tropical Rain-forest'. In Flenley, J.R. (editor), 'The Water Relations of Malesian Forests'. University of Hull, Department of Geography, *Miscellaneous Series*, 11.

Machatschek, F. (1969). *Geomorphology* (Oliver and Boyd, Edinburgh).

Money, D.C. (1972). *Climate, Soils and Vegetation* (University Tutorial Press, London).

Ollier, C. (1969). *Weathering* (Oliver and Boyd, Edinburgh).

Panton, W.P. (1964). 'The 1962 Soil Map of Malaya'. *Journal of Tropical Geography*, 18, p. 118.

Pereira, H.C. (1973). *Land Use and Water Resources* (Cambridge University Press).

Scientific American (1970). *The Biosphere* (Freeman, San Francisco).

Selby, M.J. (1971). *The Surface of the Earth*, Volume 2. *Climate, Soils and Vegetation* (Cassell, London).

Swan, S.B. St. C. (1970). 'Analysis of Residual Terrain, Johor, Southern Malaya'. *Annals Association American Geographers*, 60, p. 124.

_____(1970). 'Relationship between Regolith, Lithology and Slope in a Humid Tropical Region: Johor, Malaya'. *Transactions: Institute of British Geographers*, 51, p. 189.

Thomas, M.F. (1974). *Tropical Geomorphology: A Study of Weathering and Landform Development in Warm Climates* (Macmillan, London).

Troll, C. (1963). 'Landscape Ecology and Land Development with special reference to the Tropics'. *Journal of Tropical Geography*, 17, p. 1.

Twidale, C.R. (1971). *Structural Landforms* (The M.I.T. Press, Cambridge, Massachusetts).

Wilford, G.E. and Wall, J.R.D. (1965). 'Karst Topography in Sarawak'. *Journal of Tropical Geography*, 21, p. 44.

Wyatt-Smith, J.A. (1964). 'A Preliminary Vegetation Map of Malaya with Descriptions of the Vegetation Types'. *Journal of Tropical Geography*, 18, p. 200.

Young, A. (1972). *Slopes* (Oliver and Boyd, Edinburgh).

Index

Adiabatic rate, 32, 33; dry, 32, 33; saturated, 33.
Advection, 90.
Air masses, 65; continental tropical, 65, 71, 72; maritime tropical, 65, 71, 72.
Albedo, 38.
Angular momentum, 63.
Anticyclones, 68.
Aridity, 73, 79.
Asthenosphere, 11, 16.
Atmospheric counter-radiation, 43.
Atmospheric water-balance, 90, 108, 109.
Atmospheric run-off, 90, 91.
Atomic number, 133.
Atomic weight, 133.
Attenuation of solar beam, 40, 41.

Bernoulli's equation, 35.
Blackbody, 36.
Boundary layers, 59, 60.
Bowen ratio, 47, 48.

Capillary forces, 100, 101.
Catchment, 109.
Chemical energy, 30, 31.
Clay-humus complex, 151.
Climatic climax vegetation, 147, 158, 159.
Cloud cluster, 75.
Condensation nuclei, 94.
Conduction, 29, 30, 34, 60.
Continental drift, 16-23.
Convection, 30.
Coriolis force, 58, 59, 61, 64.
Coriolis parameter, 59, 71, 76.
Cumulus cloud, 75, 76.
Cyclic salts, 135, 136, 149.

D'arcy's law, 102.
Denudation, 11, 12.

Diffusion, 60.
Diffuse sky radiation, 38, 39, 41, 42.
Direct solar radiation, 38, 39, 41, 42.
Drainage basin, 109.
Duricrusts, 158.

Earthquakes, 16-23.
Easterly wave, 76, 77.
Ecosystems, 9, 10, 120, 147-51, 158, 160.
Eluviation, 151, 152.
Energy, 29; chemical, 30, 31; heat, 29; kinetic, 30, 35; potential, 30-5; radiation, 30.
Entropy, 7, 8, 12.
Equatorial westerlies, 75.
Evaporation pans, 95, 96.
Evapotranspiration, actual, 103-6.
Evapotranspiration, potential, 95-7, 103, 104, 149.
Exponential decay curve, 8, 12.
Extreme events, 114-16; annual series, 114-16; partial duration series, 114; probable maximum flood, 115, 116; probable maximum precipitation, 115, 116; return period, 115, 116.

Floods, 114-18.
Flow, laminar, 59.
Flow, turbulent, 59.
Fluids, 55.
Friction, 55-7; angle of, 57; co-efficient of, 56.
Fronts, 65, 66.
Frontal depressions, 66.
Frost, 50.

Geomorphological cycle, 12-15.
Geostrophic wind, 60, 61.
Glacial climate, 120.

Glacial landforms, 123-5.
Gramme molecular weight (Mole), 133.
Greenhouse effect, 42.

Hadley cell, 33, 64, 66.
Half-life, 11.
Heat, 29; latent, 29-35, 44-7, 79; sensible, 29-35, 44-7
Humus, 151, 152.
Hydration, 137.
Hydrogen ions, 134.
Hydrologic cycle, 89, 90, 102, 135.
Hydrolysis, 137.
Hydroxide ions, 134.

Illuviation, 151, 152.
Infiltration, 99-100; capacity, 99-100.
Inselbergs, 142, 143.
Interception, 97-9; loss, 97-9; stemflow, 97-9; throughfall, 97-9.
Inter-tropical convergence zone, 75.
Isostasy, 11, 12.

Jet streams, 63; sub-tropical, 63, 72, 80, 82.

Kaolinite, 138, 153.
Karst topography, 24, 25, 138, 139.
Kinetic energy, 30, 35.

Latent heat, 29-35, 44-7, 79.
Laterite, 154, 155, 158.
Latosols, 153.
Lithosphere, 11, 16.
Longitudinal stream profiles, 145.
Lysimeter, 96.

Malaysia, 20-6, 48-51, 76, 82-5, 97-100, 106, 107, 117-19, 127-30, 138-43, 145, 146, 150, 153-6, 158-60; air circulation, 82-5; climatic change, 127-9; coasts, 26; floods, 117-18; geological structure, 24-6; highlands, 24-6, 145, 146; infiltration, 100; interception, 98, 99; karst, 25, 138-42; nutrient cycling, 150; old alluvium, 26, 130; plate junctions, 20-3; potential evapotranspiration, 97; radiation, 48-51; rainfall, 76, 83-5; sea levels, 129, 130; slopes, 143, 145, 146; soils, 153-6; stream erosion, 119; temperatures, 49-51 82; vegetation, 158-6; water-balance, 106, 107.
Mars, 13, 14.
Mid-ocean ridges, 18-23.
Moon, 13.
Mor, 151, 152.
Morphometric indices, 145.
Mull, 151.

Net primary production, 147.
Net radiation, 43, 48, 49, 62, 149.
Newton's laws of motion, 55.
Nutrient flow, 149-51.

Oceans, 68-70; currents, 68-70; gyres, 69, 70; trenches, 16-23; waves, 69.
Oxidation, 137.

Pedalfer, 157.
Pediment, 144.
Pedocals, 157.
Peneplain, 12, 14, 15.
pH, 134.
Phreatic surface, 101.
Plate tectonics, 16-23.
Pluvial periods, 129.
Porosity, 102.

Potential energy, 30-5.

Radiation, 30; global, 39, 48, 49; long-wave, 36-8, 49; net, 43, 48, 49, 62, 149; short-wave, 36-8.
Radioactivity, 11.
Rainfall, 92-5; coalescence process, 94, 95; ice-crystal process, 94, 95.
Reduction, 137.
Regolith, 151.
Rift ocean, 18.
Rigid body, 55.
River flow, 109-19; base flow, 110, 111; hydraulic radius, 113, 114; hydrographs, 109-13; Manning formula, 114; s-curve hydrograph, 112, 113; sediment load, 118, 119; stage, 113; time of concentration, 111; transportation, 118, 119; unit hydrograph, 112, 113; wetted perimeter, 113.
Rocks, 15, 16; igneous, 15, 16; metamorphic, 15, 16; sedimentary, 15, 16.
Roughness length, 60.

Sahul Shelf, 23-6.
Scalar, 55.
Sea level, 129, 130; eustatic controls, 129, 130; isostatic controls, 129, 130.
Sensible heat, 29-35, 44-7.
Sheet flow, 108.
Singapore (see Malaysia).
Slopes, 142-5; debris-controlled, 142-5; gravity-controlled, 142-5; wash-controlled, 142-5.
Snow-line, 50, 121, 126.
Soils, 151; azonal, 156; brown earths, 157; chernozems, 157; chestnut, 157; gley, 157; horizons, 151, 152; intrazonal, 156; laterite, 154, 155, 158; latosols, 153; podzol, 152, 153; profile,

151, 152; serozems, 157; solonchaks, 158; solonetzs, 158; zonal, 156.
Soil moisture, 100, 101; available water content, 101, 104; field capacity, 100, 101; tension, 101; wilting point, 101.
Solar constant, 40, 43.
Solar declination, 39, 40.
Solution, 134, 135, 137.
Stefan-Boltzmann law, 36, 38, 42.
Sub-surface flow, 108.
Sub-tropical anticyclone, 63, 71, 72.
Sunda Shelf, 23-6.
Surface tension, 100.
Systems, 4; cascading, 6, 7, 102, 135; closed, 4, 5; cyclic, 8; decaying, 8; haphazardly fluctuating, 8; isolated, 4, 5, 12, 33; morphological, 6; open, 4, 5, 13, 14, 147; process-response, 7; steady-state, 5, 6, 147, 158.

Temperature, 29.
Thermal conductivity, 34.
Trade-wind inversion, 72-5.
Transpiration, 95.
Tritium, 91.
Tropical cyclone, 78, 79, 82.
Tropical depression, 78.
Tropical rain forest, 46, 47, 98, 99, 104.
Tropical storm, 78.

Unloading, 142.

Vector, 55.
Vertical motion, 93, 94; convective, 93, 94; banded, 93, 94.
Viscosity, 57.

Weathering, 135-7, 149, 153.
Wien displacement law, 36, 38, 42.
Würm (Wisconsin) glaciation, 128-9.